Designing Creative Tools

Revealing Principles Behind Tools That Shape Art, Culture and Imagination

Couche, Guillaume
Rai, Sanya
Shackleton, Richard

Designing Creative Tools: Revealing Principles Behind Tools That Shape Art, Culture and Imagination

BIS Publishers
Timorplein 46
1094 CC Amsterdam
The Netherlands

bis@bispublishers.com
www.bispublishers.com

ISBN 978-90-636-9776-1

Designing Creative Tools

Revealing Principles Behind Tools That Shape Art, Culture and Imagination

Guillaume Couche
Sanya Rai
Richard Shackleton

BIS PUBLISHERS

Contents

Foreword

There's an old saying that necessity is the mother of invention. That is perhaps nowhere more evident than in the way creative people use tools in their work. As a self-taught artist, I was never taught the correct tools to use and consequently tended to learn through guesswork, experimentation, and by accident. Many artists have evolved styles based on the tools they have had easy access to. I don't know which painter was the first to paint directly using the palette knife — a tool intended for mixing paint and cleaning the palette — but it has certainly become a familiar alternative to the brush for quickly applying larger areas of flat colour to the canvas.

Similarly, when starting out, I couldn't afford to keep buying new brushes, and this felt like a big disadvantage. As it was the small, delicate ones that would wear out most quickly, I ended up having to use the clumsier, larger, or worn ones, which made reproducing fine detail much harder. This actually turned out to have benefits, as it forced me to innovate, find alternative ways of suggesting forms, and adopt a looser, less representational style. It also meant the painting process was much quicker.

The 19th-century innovation of collapsible paint tubes was primarily a way of speeding up the painter's process by removing the need to mix pigments into workable colours manually. It also had the unanticipated advantage of freeing artists to take their work out of the studio for the first time, providing the means to depict landscapes in the open air; Impressionism was born as a result. This pattern of tool design shifts leading to creative liberation is a constant in art history. The invention of the camera was a major disruptor, freeing painters from the responsibility of recording literal appearances and allowing them to express new stories. Once those constraints were gone, artists experimented with abstracted lines and surreal worlds. Similarly, Warhol took to commercial screen printing to mass-produce his art, showing how creative people often repurpose tools from other fields to find original paths.

I've always learned more from talking to artists across different disciplines about their tools. That's how I discovered that lighting is a key tool for portraiture: learning to sculpt with light from cinematographers showed me how tools can influence and change the way you think about what you're trying to achieve. That's what a good tool can do: it doesn't just make your job easier, it increases your understanding of the craft and the different ways you can affect the outcome.

Beyond traditional methods, staying curious about new ways of doing things provides a sense of wonder vital for creative growth. The excitement that comes with discovering a new tool is one of the best feelings, opening up fresh worlds of possibility. These breakthroughs do not just offer convenience; they fundamentally alter our perspective and understanding. Yet, despite modern options, there is something in us that likes engaging physically to have an effect. We are hardwired to make things, finding a unique satisfaction in physical interaction. As humans, we feel a deep contentment in objects that remain after the work is done. The tools we use are part of our evolution, and a drive to create with them is exactly why we have evolved as a species.

— Jonathan Yeo, February 2026

► *HM King Charles III by Jonathan Yeo*
Oil On Canvas (2024)

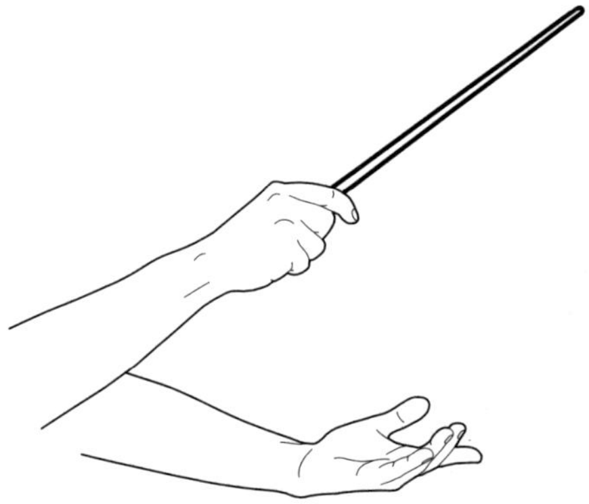

Introduction

Although the exact definition of culture has been the subject of much discussion since the inception of anthropology, most agree that it is not a monolithic concept. In the 1950s, American anthropologist and philosopher David Bidney and English evolutionary biologist Julian Huxley popularised a segmentation of culture into *artefacts* (the material things people make), *sociofacts* (social structures such as families and governments), and *mentifacts* (the ideas, values, and beliefs that we share).[1] While many would argue that culture, considered as a whole, is emergent rather than a deliberate project, artefacts, sociofacts, and mentifacts are all products of what we create — consciously or not. Everyday objects, works of art, houses, and temples are all created by humans, directly or indirectly. Folklore, social structures, and governments are also created. The ideas in our minds are constructed, generally using the language we speak and often building on the ideas of others that have reached us through education, poetry, and shared stories.

However, to be fully understood, these cannot be considered in isolation, as an interplay exists between them. Typically, the artefacts we create reflect our mentifacts and sociofacts. This is evident throughout the long history of religious architecture, from the stone pillars of Neolithic Göbekli Tepe to the Egyptian pyramids and Gothic cathedrals. Our sociofacts also reflect our mentifacts. Political systems, for instance, are often preceded by philosophical ideas — good and bad. For this to happen, there is always an *object*, a *system*, or a *method* sitting at the centre of the creation process: intuitions become theories when articulated using agreed-upon argumentative structures; a group of like-minded individuals becomes a political party when publishing a manifesto or a programme; an idea becomes a design when traced on paper using a pencil. These are, in effect, tools and among the various artefacts, mentifacts, and sociofacts of human culture, they may play the most important role, as they determine the creation of everything else.

Tools can be material or immaterial, simple or incredibly complex, familiar or specialised, hardly noticeable or obviously in the way. They enable the creation of things that were simply not possible to be created before they existed. The hand axe enabled the built habitat; the camera, the capture on paper of what we see; 3D printing, the materialisation of impossible geometries. However, tools determine what we create, not only by what they enable but, also, by the restrictions they impose. The chisel led to the idea of finding a shape within a block of stone through a process in which actions are subtractive, which, in turn, seems to have inspired[2] Aristotle's concept of potentiality.[*] Perspective compelled painters to adopt a single viewpoint and adhere to projection rules, rather than the freedom of symbolic representation, leading to more consistent representations of depth from the Renaissance onward. The fixed-key system on the harpsichord, and later the piano, restricted composition to specific pitches and harmonies now associated with Western music history, from the classical fugue to the endlessly reused four chords of most pop music hits.

But what remains perhaps the biggest, and yet least tangible, influence of tools is how they inspire us. The metaphor, for instance — by suggesting one thing, that may be too abstract, can be understood through the description of another, more familiar one — inspired both the conceptualisation of complex ideas by philosophers and the design of the user interfaces through which we use our computers and smartphones. As for physical tools, archaeologists have found ample evidence in most human societies of 'mill songs' — ritual work songs inspired by the characteristic sounds and rhythms of grinding tools, likely to be one of the earliest forms of shared musical experiences. Fast-forward to the 1980s: the popularisation of the rhythmic grid on drum machines such as the Roland TR-808 inspired the precise, repetitive patterns — well-suited to these machines — that are now common across almost all popular electronic music genres, including techno and house. Finally, one needs only to listen to an artist speak about the materials or physicality of their tools to understand that humans form a bond with them that runs deeper than a *strictly work relationship*.

The proposition that tools dictate what we create may be taken as a form of technological determinism, or, put simply, a denial of human creative agency. This misconception stems from a common confusion between tools and technology, leading to arguments built on shaky foundations and reasoning that is fundamentally flawed. Technology is the practical application of scientific knowledge. As such, it may appear to evolve in one direction and at a somewhat predictable pace, as proposed by Moore's Law[†3] — in fact, more of an observation focused on one very specific area of technological development, and one whose relevance today is best approached with caution. A tool, on the other hand, is a specialised object designed to facilitate a task. While the development of

[*] The latent possibility within an object to achieve a specific end (telos) or state of being. Aristotle distinguishes between 'natural' potential — which realizes itself given the right conditions — and 'accidental' change, which occurs outside a thing's inherent purpose.

[†] In 1965, Gordon Moore, the co-founder of Fairchild Semiconductor and Intel, observed that the number of transistors in an integrated circuit doubled approximately every two years, shaping the way many commentators and decision-makers have predicted the future of the computer industry in the last decades.

technology expands the horizons of what is physically achievable, the design of tools proceeds from individual, granular, and often divergent decisions on how to apply this potential. Hence, focusing on the tool does not remove the human from the equation; rather, it considers the human twice: when we use the tool and, before that, when we *design* it.

In 2001, British painter David Hockney published a book intended to reveal the secret of the great masters, including their alleged use of optical aids to create better perspectives and light rendition.[4] It drew vigorous criticism, its most ardent opponents pointing to a lack of evidence.[5] Whether or not he was right is, in some regards, less interesting than what it would have meant to the art world, a sentiment famously summarised by American cultural critic Susan Sontag the same year: 'If David Hockney's thesis is correct, it would be like finding out that all the great lovers of history have been using Viagra.'[6]

Interestingly, nobody seems to have a problem with painters using brushes — rather than their bare hands, as their predecessors did. To some critics, Van Gogh's adventurous use of pigments makes him 'the greatest colourist ever,' even if that means we can no longer see his work in the colours he intended.[7] Yet, the very idea that a painter may have used a lens for a quick outline, or a small mirror to check a colour, seems sufficient to render them unworthy of their status as an artist.

But what are pigments, paintbrushes, lenses, and mirrors if not one and the same thing: tools to achieve a creative end? Most would agree that artists are more than the sum of their know-how, tricks, and techniques, including those mediated by the use of tools. However, if we refuse to acknowledge, discuss, and understand the role of the tool in the creative process, we risk ascribing to it a greater role than it warrants, blinded by reasoning akin to a

'focusing on the tool . . . considers the human twice: when we use the tool and, before that, when we design it.'

tautology: 'tools cannot be too important, otherwise they would become too important.'

What the David Hockney story shows is that without a critical examination of the tools, we are forced to draw a line between what is acceptable and what is not. Brushes and pigments, one imagines, are old and *traditional* enough that they get a pass. Lenses and mirrors, deemed too *high-tech*, do not. But where to draw the line? How old does a tool need to be to be considered *traditional*? The recurring theme of the *technological* supposedly opposed to the *traditional* shows a deeply rooted fear.

Already in 1958, French philosopher Gilbert Simondon, in his essay *On the Mode of Existence of Technical Objects*, declared, 'Culture has become a system of defence designed to safeguard man from technics. This is the result of the assumption that technical objects contain no human reality.'[8] This is not to say that this fear has no foundation. In fact, one only needs to compare the two ends of the spectrum to understand the dilemma we are faced with today: How can we reconcile ourselves with the idea that an image generation chatbot and a paintbrush are both *tools*, when generating a painting with a prompt has so little in common with *actually* painting it, stroke by stroke, on canvas?

Examining tools does not imply that all tools are equal; rather, quite the opposite: it is only by acknowledging tools as bearers of unique properties that we can understand how they affect the creative process and avoid the pitfalls of extreme relativism. Examining tools is an attempt to answer why we find the use of some tools more virtuous than others; why some seem to magnify creative abilities while others feel like cheating and leave the user with a sense of emptiness. Examining tools involves drawing lessons from all of that — lessons that can inform the design of new tools, which, ideally, will allow for the further enrichment of our culture.

Why This Book, This Way

With writing comes organising ideas and, at times, building theories. When a theme has been largely unexplored, there may also be value in presenting knowledge in a manner that allows the reader to form their own analysis and pick insights the authors may have overlooked, had they processed the raw material through their own lens. This is why nearly a quarter of the words in this book are from practitioners, even though the questions are from the authors, which certainly provides direction for the interviews. Moreover, half of this book is a curated list of iconic tools, each of which balances factual information with design analysis. Beyond the obvious illustrative purpose of that chapter, it felt crucial to show the incredible diversity of creative tools in a format where Photoshop, renowned for image retouching, can sit right next to a Fender guitar, which revolutionised modern music, a few pages away from the Singer Model 15, a sewing machine that democratised fashion at the turn of the 20th century.

This will allow readers to form new associations by presenting an alternative narrative to the dominant ideas that make it easier to associate Photoshop with other software applications, such as a web browser, even though their functions are fundamentally different — one intended to *create*, the other to *consume*. This book posits that years of thinking through *technological proximity* have led designers to apply design principles originally developed for mass consumerism to creative tools, thereby failing to propose a divergent voice in the process. This may carry dire consequences for tomorrow: the automation of tools, the commodification of creation, and ultimately the flattening of culture — all driven by economic imperatives.

Creative Tools?

Finding the right title required much thought. Indeed, to cover everything from paint tubes to cameras, musical instruments, and software applications, it had to convey a sense of diversity. But why *tools* and not *instruments*, *devices*, or simply *products*?

If, at first glance, the term *tool* may seem plain, its neutrality is its greatest asset. By comparison, *instrument* is only really used in a creative context with music. Similarly, a *device* is typically a small connected machine, such as a smartphone or smartwatch — not a fitting description for a piano or a paint tube. Speaking of machines, musical instruments with some form of automation have been referred to as *machines* in the past: *drum machines* come to mind. Yet, a *machine* is more likely to be understood as large and industrial; furthermore, not everything in this book is *technically* a machine.

This leads us to the ever-approachable term *product*, which resonates with most designers and could have been chosen for convenience. However, as further developed in Chapter 1, the very definition of *product* insists on the object as a produced entity rather than its role.

The very definition of *tool* is linked to the idea of creation — or destruction. It is straightforward and honest, notably in its uncompromising role of function. Even Adobe, which admittedly has sufficient communication staff to brainstorm such questions, consistently returns to *creative tools* when referring to its applications. There would be something to be said about choosing *creative tools* over *creation tools*. After all, it is not the tools themselves that are creative! But *creation* carries a heavy, almost industrial, and sometimes even theological weight. As for the most unusual gerund *creating tools*, beyond the tongue-in-cheek wordplay for the cover, it would have been a nightmare to disambiguate throughout the book.

Ultimately, a quick search confirms that *creative tools* is the most commonly used expression, a cruel irony in a context where many fear that artificial intelligence (AI) is placing human creation in jeopardy. At the very least, it forces us to contemplate what may be one of the most important questions ahead of us: what if tools, not humans, were to become the ones being creative? As a meagre consolation, it seems that, for as long as we speak of *tools*, their role will be confined to that of intermediaries between human intentions and tangible outcomes. Let's just hope that no one ever has to write a book about *creative entities*.

'years of thinking through technological proximity have led designers to apply design principles originally developed for mass consumerism to creative tools . . . failing to propose a divergent voice in the process.

This may carry dire consequences for tomorrow: the automation of tools, the commodification of creation, and ultimately the flattening of culture — all driven by economic imperatives.'

Users, Creators, and Artists

'Time was when the mystery and wonder of handicrafts were well acknowledged by the world, when imagination and fancy mingled with all things made by man; and in those days all handicraftsmen were artists, as we should now call them.'

— William Morris, 1882[9]

To avoid overreliance on the term *user*, those who use creative tools are often referred to as *creators* in this book; this should be understood in the widest and most inclusive sense. Hence, the term *creator* encompasses all humans who create, whether they are celebrated artists, emerging designers, experienced craftspeople, amateur enthusiasts, or anything in between. To be clear, this is certainly *not* a reference to *content creators*, even if social media companies are trying hard to pre-empt the term *creator* to mean just that — as illustrated by the 2021 Meta rebrand event, where Mark Zuckerberg and his management team mentioned 'creator' and 'creators' almost 50 times,[10] more than once every other minute. Yet, it is clear from the presentation that their definition of the term 'creator' is much narrower than its standard usage in English. Specifically, they refer to the individuals whom they hope will build digital assets and businesses in their ecosystem. For maximum clarity, *content creator* is used in this book only to refer to people who create content such as videos, images, and writing for the internet, usually monetised through advertising. This book does not dispute that content creation is a form of creation; simply that reducing creation to content creation is, in the long run, dangerous.

Finally, the terms *artists*, *designers*, *writers*, *painters*, and *musicians* are used occasionally to add life to the text, never to differentiate them from other creators.

‘what if tools, not humans, were to become the ones being creative?

. . . reducing creation to content creation is, in the long run, dangerous.’

[1] Stephen Mitchell, ‘Continuity: Folklore’s Problem Child?’, in Folklore in Old Norse — Old Norse in *Folklore*, ed. Daniel Sävborg and Karen Bek-Pedersen (Tartu: University of Tartu Press, 2014), 47.

[2] S. Marc Cohen and C. D. C. Reeve, ‘Aristotle’s Metaphysics’, in *The Stanford Encyclopedia of Philosophy*, ed. Edward N. Zalta and Uri Nodelman, Spring 2025 ed. (Metaphysics Research Lab, Stanford University, 2025), plato.stanford.edu/archives/spr2025/entries/aristotle-metaphysics/, sec. 12.

[3] Gordon Moore, ‘Cramming More Components onto Integrated Circuits’, *Electronics* 38, no. 8 (April 19, 1965).

[4] David Hockney, *Secret Knowledge: Rediscovering the Lost Techniques of the Old Masters* (New York: Viking Studio, 2001).

[5] Sven Dupré, ‘Introduction. The Hockney-Falco Thesis: Constraints and Opportunities’, *Early Science and Medicine* 10, no. 2 (2005): 125–36.

[6] Paul Lieberman, ‘Artistic Fact or Optical Delusion?’, *Los Angeles Times*, December 3, 2001, latimes.com/archives/la-xpm-2001-dec-03-mn-10977-story.html.

[7] Jonathan Jones, ‘Van Gogh’s True Colours Were Originally Even Brighter’, *The Guardian*, May 2, 2013, theguardian.com/artanddesign/jonathanjonesblog/2013/may/02/van-gogh-original-colours-brighter.

[8] Gilbert Simondon, *On the Mode of Existence of Technical Objects*, trans. Ninian Mellamphy (Paris: Aubier, Editions Montaigne, 1958; repr., University of Western Ontario, 1980), 1.

[9] William Morris, *Hopes and Fears for Art* (Cambridge: University Press, John Wilson and Son, 1882), 9–10.

[10] Meta (Facebook) ‘Connect 2021 Metaverse Event Transcript’, *Rev*, October 28, 2021, rev.com/transcripts/meta-facebook-connect-2021-metaverse-event-transcript, accessed January 19, 2026.

1 | Creative Tools & Design

Creation knows no boundaries and encompasses a vast array of human activities, from imagining fictional worlds and mythical fables to reproducing the beauty of nature and attempting to capture the very meaning of our life on Earth. Whether an object or a performance, the work created will likely outlive the creator, and a lifetime could be spent discovering the work of others.

American author Randall Munroe, famous for providing 'serious scientific answers to absurd hypothetical questions,'[1] estimates that the date after which there were simply too many books to be able to read them all in a lifetime happened around 1500 — if we consider only those written in English.[2] At today's publishing rate, a lifetime would not be nearly enough to read a small percentage of the 200,000 books published each year — in the UK alone.[3] Obviously, books are only one example of the many things humans create: there are also movies, plays, paintings, songs, video games, dress patterns, sculptures, websites, choreographies, light shows, murals, podcasts, photographs, toys, collages, installations, remixes, designs of furniture, buildings, logos, and the list goes on.

Creation is diverse — as diverse, in fact, as our ideas of what may be created. Yet this is not, on its own, what makes it uniquely interesting; rather, the reason *why* we create does. To fully appreciate that, we first need to examine other human activities.

Communication happens in order to share information, ideas, and feelings; studies and training are needed to learn, acquire skills, and become better at what we do; work is a way to earn a living, be useful to society, and change the world; going to the gym is a way to maintain good health and stay in good shape; consuming allows us to relax, have fun, or just find pleasure. In *Creativity*, Hungarian-American psychologist Mihály Csíkszentmihályi proposes that most human activities are *exotelic*, a Greek term for what is externally motivated.[4] In his own words: 'We do them not because we enjoy them but in order to get at some later goal.'[5] He points out that only two forms of human activities differ from the rest in that regard: competitive sport and creation. He describes those as *autotelic* activities and goes on explaining that an autotelic activity is 'something that is an end in itself.'[6] But then, what about people who create because they've been asked to do so? Certainly, designers cannot have a career without clients, and fine artists often work on a commission basis. And what about those who create in order to be famous or — at least — acknowledged?

Working for a client, producing a new album for a record label, or finishing a design project to earn a university degree is not mutually exclusive with loving what one does. In fact, most great artists, designers, and craftspeople are likely first and foremost motivated by the act of creation itself; they *then* use external goals as socially acceptable excuses to pursue their passion. In that respect, it may sometimes be easier to be a professional than a hobbyist, for it provides a remarkably effective cover story for how one chooses to spend their time.

Still, the direction creation has taken in recent years seems less romantic than the vision proposed by Csíkszentmihályi: after all, isn't the whole 'content creator' phenomenon a *firework* of the *exotelic* and the ultimate proof that creation is turning into something entirely motivated by the most prosaic considerations?

Perhaps it is — or maybe we are mistaking consequences for causes: we may already be witnessing the nefarious effects of having failed to consider creation as a distinct human activity, in a context of rapid technological change that has fuelled the equally rapid evolution of creative tools. In many cases, this occurred without any notable intervention from the very people who should have been leading conversations or — at least — asking questions: designers.

The Problem with Design

To fully appreciate the lack of attention given to creative tools in design, one first needs to consider the underlying structure of any disciplinary discourse: language.

Of all the titles in the design industry, *product designer* is amongst the most common, and at a superficial level, *product* sounds vague enough to allow us to imagine a vast array of possibilities. In fact, there are product designers who design furniture, product designers who design everyday objects, and, increasingly, product designers who design non-physical things, such as web and mobile applications.

But the term *product* is not as neutral as it may seem. The definition of a product is 'something produced'*[7] or, said differently, the end link of a chain rather than a connecting one. Of course, nothing prevents something produced from being used to produce something else — tools, for one, have to be produced, at some point. Yet, calling everything a *product* already assumes a stance where we view the designed object not for what it might enable, but as the end result of a creative process understood solely as existing within the boundaries of the designer's work.

It is even more telling to consider the term ascribed to the people for whom things are

---

* The online Cambridge Dictionary defines 'product' as 'something that is made to be sold, usually something that is produced by an industrial process or, less commonly, something that is grown or obtained through farming'.

designed: *users* — literally those who use. In fact, there is no easy alternative to the verb *using*,* which, stemming from Old French, remains a close synonym of *consuming*.

In effect, before the designer even begins work, it is not only the object of design which is denied its potential to exist beyond its status as a product; the interaction it may mediate is, by default, defined as consumption.

Outside the design studio, this is echoed in the expression *consumer electronics*, used as an umbrella term for all objects that incorporate some form of electronics. Today, it can refer to almost anything and, in fact, some of the biggest manufacturers of creative tools announce their new releases at CES (the Consumer Electronics Show). It could be argued that this is a legacy name and that nobody truly pays attention to the meaning of acronyms. Yet it perpetuates a symbolic equivalence between, on the one hand, refrigerators, games consoles, and anything designed precisely for consumption, and, on the other, creative tools. Furthermore, this symbolic equivalence is reinforced by a shared design language and, often, similar technological architectures. After all, is a professional digital drawing tablet, such as a Wacom Cintiq, that different-looking from a tablet used to watch videos? Is a security camera not using an image sensor like a photography camera? And what about the laptop? Some may use theirs to play video games, others to type a report or design a logo, and some may do all of the above.

This raises the larger question of the digitisation and hybridisation of all things,† both of which have hindered the construction of a design discourse specific to creative tools, at a time when design, in its entirety, had to be rethought.

With the generalisation of electronics and computing across nearly all objects, new questions have emerged, gradually reshaping what it means to design. At the centre of such questions is the behaviour of objects and specifically how humans interact with them. Interaction design, which seeks to address these questions, has been one of the most important additions to the design corpus since the late 1980s. The core idea is to shift the designer's focus from the object itself to the space that separates it from the person using it — the space where interactions occur. Although developed primarily for software, this idea has long extended to physical interactive objects as well.

In his book *Designing Interactions*, British designer William Grant 'Bill' Moggridge provided an extensive overview of the field at the turn of the millennium.[8] Widely acclaimed upon publication, it remains, two decades later, one of the most influential books on interaction design. Not only does it provide a history of how interaction design emerged, but it also features 40 interviews with design practitioners, engineers, and entrepreneurs, organised around case studies such as the development of the first Apple iPod. As for the author himself, he designed the first laptop computer and co-founded IDEO, one of the most successful design consultancies of the last 40 years.

Perhaps due to the young age of the ideas it promoted at the time of writing,‡ *Designing*

* Despite efforts to remain consistent with the argument by avoiding these terms, the continued necessity of employing 'users', 'using', and 'products' for the sake of clarity throughout this book is a testament to how deeply ingrained they are in the English design vocabulary.

† Sometimes described as technological convergence: the process by which previously distinct technologies — such as telephony, computing, and media — merge, driven by standardisation and digitisation.

‡ *Designing Interactions* was first published in 2006.

Interactions did not clearly distinguish creation from other activities that the devices and software it studied tended to.* And there was a certain logic to that: what transpired from the book's very structure was the building of knowledge through the recording and analysis of field experience. Its goal, it appears, was to outline principles that could apply widely, rather than risking to fragment a nascent discipline that was yet to find its place among more established ones, such as industrial design or architecture.†

However, there might have been another reason for not acknowledging creation as a distinct activity that would have called for a different design approach: as stated earlier, interaction design was born from the rise of the personal computer and software — as objects of design. Yet what distinguishes computers from most tools humans had designed before is their open-ended nature, which leaves activity-specific tasks to software applications. This is why the term *platforms* is so often used, a concept that has gained popularity, not least because of the huge business opportunities modelled on it.‡ For most people, the introduction of the personal computer, as a platform, came with the suggestion that they could move from using a myriad of tools such as typing machines, cameras, and musical instruments, all of which were specialised, to using a single multifunctional one. Today's computers and smartphones can be used effectively for both creating and consuming.

The consequences of this de facto hybridisation of tools that support a wide spectrum of human activities are wide-ranging and far-reaching. With multifunctionality came multitasking, and it is therefore not entirely surprising that the pioneers of interaction design preferred to ignore the specifics of segmented activities — which may have looked like an unnecessary legacy of the past — to focus on more universal principles. One such principle was proposed by David Liddle, one of the designers of the Xerox Star, the first personal computer. Interviewed in *Designing Interactions*, he explained that how technology — and in fact any product — is used follows three phases of development and adoption, named after what he described as three distinct markets: enthusiasts, who, beyond a product, want to explore new technologies for the sake of it; professionals, who seek utility and the maximisation of productivity; and consumers, who seek lifestyle fit and automation for effortless integration into their daily lives.[9]

This is an interesting theory that follows an indisputable market-scale logic and, 20 years on, appears to have proven correct in the author's field. Apple, with the Apple II in 1977, the Apple Macintosh in 1984, the Apple iPhone in 2007, and the Apple Watch in 2015, exemplifies the shift from enthusiasts to professionals and eventually consumers.§[10] But is that really the

* While interviews like Cordell Ratzlaff's explicitly contrasted 'creating content' with consumption, and others noted the dual role of MacPaint as both tool and toy or The Sims as a game centred on creation, the book generally integrated these into a continuum of interaction rather than treating creation as a singular, isolated category.

† By way of comparison, De Architectura by the Roman architect Marcus Vitruvius Pollio — generally regarded as the first known work on design theory in the Western world — was likely written between 30 and 20 BCE.

‡ The profit share generated by third party software applications running on the Apple iPhone and directly benefiting Apple — as provider of the platform — is probably one of the most studied and commented business cases of the last two decades.

§ This is not achieved solely through the introduction of new products. In *Designing Interactions*, Cordell Ratzlaff — another interviewee who led the Human Interface Group at Apple from 1990 to 1999 — noted the transition underway at the time, stating that 'the whole application/document model was developed for people who were creating content' and contrasting it with what he saw as a shift towards pure content consumption.

only desirable pathway for everything that is designed?

'[in the consumer phase], the priorities for the product have dramatically changed, and one thing that we always see is that most of the important controls become automatic.

Today you buy a 35mm camera . . . it will read the film speed automatically, and set the exposure automatically, and set the flash automatically, and everything is done, and actually a chimpanzee can take a pretty good photograph with it.'

— David Liddle, 2003[11]

In AD 100, disappointed by how Rome was changing, Roman satirical poet Juvenal famously denounced a society only preoccupied by 'bread and games.'[12] Under the progression model proposed by Liddle and promoted in the foreword to *Designing Interactions*,[13] consumption and recreation appear as the ultimate offerings of design, led by market size, opportunity, and cost reduction. Any other activity that new tools might support — unassisted creation, for instance — becomes part of the anecdotal and, in fact, the transitional. Freedom is traded for convenience; human creation remains viable primarily on the condition that it is automated and entertaining.

It is tempting to think of interaction design as its own, somewhat niche discipline, focusing on personal computers, software applications, and the occasional interactive installation. In reality, at a time when entire industries, such as automotive, are shifting their focus toward questions of *experience* and *autonomy*, it becomes clear that the boundaries of interaction design have increasingly blurred, with ramifications that extend far into other disciplines — from industrial design, where most designers today primarily focus on smart devices, to architecture, where new ideas such as *responsive building envelopes*, *sentient habitats*, and *spatial wayfinding* denote a quiet revolution that contrasts with the field's expected conservatism.*

Initially coined by design researcher Brenda Laurel,[14] *User Experience* — now often abbreviated to *UX* — has become widely used and is frequently equated with design itself, in a context where UX-related roles are projected to be among the fastest-growing segments in the field.[15] This is thanks, in no small part, to what is arguably one of the most influential design books of all time: *The Design of Everyday Things* by American researcher Donald Arthur 'Don' Norman.† In it, Norman articulates what have become the tenets of UX, interweaving them closely with Human-Centred Design — now frequently referred to simply as HCD.‡ This methodology posits that successful design must stem from a deep understanding of human needs and behaviours, rather than merely adhering to a technical specification.

It is a cruel irony that the very ingredients that led to the success of UX — and more generally the HCD perspective — are also what make these approaches so ill-adapted to the design of creative tools. Indeed, by focusing on the broader goal of increasing access to technological products, HCD ultimately promoted a particular idea of design's role as a

* We are not referring here to political ideals, but to architecture's — perhaps undue — reputation for being stuck in the past, given its heritage as one of the oldest theorised design disciplines.

† While true influence is difficult to quantify, *The Design of Everyday Things* is a staple on design reading lists and, as of January 2026, remains by far the most-rated book in the 'Design' category on Goodreads.

‡ In the 2013 edition of *The Design of Everyday Things*, Don Norman points out that neither 'User Experience' nor 'Human-Centred Design' was used in the original 1988 edition. However, the spirit was certainly there. Norman is often referred to as the 'godfather of UX', and as he put it regarding the first edition: 'The entire book was about HCD.'

way to ensure tools adapt to the user and that *simpler* always means *better*.

Interestingly, in a 2005 article titled 'Human-Centred Design Considered Harmful,' Don Norman pointed out the limits of the theory he had helped promote. In the article, he observes that 'some of the most successful objects of our modern, technological world [...] have been designed without the benefit of user studies and the methods of human-centred design.'[16] In a simple demonstration, he shows that there are countless designs proving that humans can, in fact, successfully adapt to technology. What do the examples he picks have in common? Aside from the automobile and sporting equipment, they are all tools — primarily used to support creative activities such as writing, making music, taking photographs, and woodworking. Why 'harmful', then? After all, he could have titled his article 'Human-Centred Design Considered Not-That-Useful.' His most serious concern is that 'focusing upon humans detracts from support for the activities themselves'[17] and therefore proposes that we should shift to what he calls 'Activity-Centred Design.' But the year is 2005: Human-Centred Design has proven itself as a way for designers to get a seat at the table and help *technology* companies conquer the world by making products that appeal to the masses. Within a year, Apple would release the iPhone and, by doing so, change the meaning of the personal computer* forever.

More recently, Don Norman has reiterated his criticism of HCD. In *Design for a Better World*, published in 2023, he advocates for a broader perspective, arguing that HCD fails to account for societal questions and environmental ecosystems.[18] But how do you make an entire generation of designers change their perspective and unlearn design principles that are largely taken for granted? In effect, the popularity of HCD likely has less to do with elaborate design arguments than with its status as a common sense alibi that translates complex human needs into a simple, market-ready slogan: 'Make it easy, don't frustrate the user, ship it.'

In *Thinking, Fast and Slow*, Daniel Kahneman proposes that once we accept a theory, it becomes 'extraordinarily difficult to notice its flaws,' even if observation contradicts our belief.[19] In other words, even if HCD — as understood by most designers — is ill-suited for the specificities of creative tools, it remains the dominant, one-size-fits-all solution that conveniently aligns with the vision of certain companies for whom a creator is, at most, a glorified consumer.

From Theory to Practice?

If both the language of design and its most popular ideas seem unsuited to creative tools, it would be wrong to conclude that this has resulted in a lack of good design. In fact — and hopefully Chapter 3 serves as a testament to this claim — we live in a time when great designs of creative tools are widely available and, for the most part, affordable as well. One possible explanation is that the idea of designing creative tools has attracted some of the greatest minds of their time. Indeed, behind the most iconic of these tools, we find brilliant inventors and entrepreneurs — and yet, few who would be

---

* *Personal* is often understood in opposition to *professional*. In fact, the term 'personal computer' was coined to refer to a new paradigm where computers would no longer be shared — as they had been at the dawn of the computing era. This vision only truly became a reality at scale with the smartphone. Consequently, the societal effects of the smartphone can be analysed not merely through the technological lens of mobility or connectivity, but by focusing on one fundamental question: What happens when humans establish a one-to-one relationship with a computer?

'Human-Centred Design . . . translates complex human needs into a simple, market-ready slogan: "Make it easy, don't frustrate the user, ship it."'

considered, let alone formally trained, as designers.*

Architects speak of *vernacular architecture* for buildings designed outside of the academic tradition — or, as English academic Ronald Brunskill put it bluntly, 'designed by an amateur without any training in design.'[20]

If vernacular architecture — estimated[21] to constitute 80% to 98% of the world's built environment — may seem to miss the sophistication of 'proper' architecture, it is often lauded for being more effective in its use of materials and more adapted to its environment.[22] This stems from its nature as a dynamic design system integrating multiple factors — including physical, cultural, and economic ones — which can be compared to a Darwinian evolutionary process, famously summarised by English polymath Herbert Spencer in 1864: 'survival of the fittest implies multiplication of the fittest.'[23]

It serves as a great reminder that good design can happen without design theory — or at least without *appropriate* design theory. There is, however, a possibly less provocative explanation for the emergence of great creative tools — one that has less to do with the *how* and more to do with the *why*.

* In the curated list of iconic tools in Chapter 3, very few of the individuals mentioned were professionally trained as designers. However, it is worth noting that for the most recent examples, particularly in software, the development teams likely include designers — though in these instances, it is impossible to know precisely what role they play in truly defining the tool.

A common motivator among those who design creative tools is that they are often themselves end users of the tools. As a consequence, even if they miss out on certain abstract design concepts, methodologies, or workflows, they possess, at their core, something infinitely more valuable to anyone who designs: *insights*.

German inventor and photographer Oskar Barnack designed the first Leica because he instinctively knew the creative potential of a camera he could carry everywhere.† American instrument designer Roger Linn changed the course of music history because, as a session musician in his youth, he understood the possibilities of a portable device on which one could compose natural-sounding drum loops.‡ With Unreal Engine, American game designer Tim Sweeney revolutionised how video games are made because he foresaw that a good game engine would enable him, and many others, to create not one but an array of different video games.§

Finally — and this will become a crucial element as we look ahead — all of this has been made possible by the increased accessibility of prototyping and production methods, illustrated in recent years by the emergence of affordable and reliable 3D printing. This, combined with DIY electronics and widely available software development tools,¶ allows — in theory — anyone to make almost anything. The experiences of British artist Jonathan Yeo# and Roger Linn** show how individuals and small

† See the Leica 1 on pages 94-95.

‡ See the Roger Linn Akai MPC Series on pages 122-123.

§ See Unreal Engine on pages 128-129.

¶ To this could be added AI in the form of coding agents, even if the novelty of the phenomenon makes it harder to assess, and its long-term effect on design might be a net negative — as explored later in this section.

See the interview on pages 66-71.

** See the interview on pages 52-58.

teams can now design, build, use, and sometimes distribute complex creative tools in ways that would have been unimaginable just two decades ago.

If It Isn't Broken, Why Fix It?

Design, as a discipline, has so far failed to understand creative tools as distinct objects of design, yet they appear to have done just fine. However, there are several reasons why continuing to ignore this blind spot may be unwise — and perhaps even dangerous for culture at large.

We lose most things because we fail to organise them. The ease of the digital copy has led us to assume that this can be remedied by simply duplicating everything in the cloud. This trend has a tangible effect on the planet[24] — yet, once we forget our archived photos, aren't they lost to us anyway? Similarly, if we archive great creative tools in museum storage, will we really learn from them? Isn't *archiving* just a polite term for *disposing of*?

Building knowledge is not about ignoring what already exists and constructing theories out of thin air. Rather, it is about curating what has been done, formulating principles, and organising ideas, so that the very essence of what humanity has learned together can be shared and benefit those who follow. Sometimes, this happens on a local scale; for instance, in organisations like Adobe — the largest software company in the creative tools industry — where hundreds of designers are reported to work.[25] However, internal guidelines and design best practices are often better suited to preserving established approaches than to opening new perspectives.

Without design theories aimed at building transversal knowledge, designers who lack a profound, innate grasp of the creative activity at hand are condemned to reproduce the same cycle: either replicating established archetypes or applying principles centred on the user-consumer rather than the user-creator. This, in turn, may lead them to focus on the wrong priorities; for instance, the automation of everything in search of providing a satisfying, albeit bland, *experience of creativity*.

Moreover, even as users of their own creative tools, the depth of insight possessed by great designers often stems from an outsider's perspective. Jonathan Yeo explores the potential of sculpting tools through the eyes of a painter-collagist;[*] American writer Natalia Illyin examines writing tools from the standpoint of a graphic designer;[†] and Norwegian designer Mats Herding Solberg brought his unique relationship with the notebook to the reMarkable project,[‡] drawing on his background as an industrial designer. Yet, over time, the hybridisation of tools narrows the possibilities for individuals to gain truly divergent experiences, and for differing ideas, perspectives, and practices to collide and progress together — a key prerequisite for healthy design, vernacular or otherwise. This reinforces the need for a design theory that encourages outward-looking perspectives on the creative tool.

Finally, while the increased accessibility of prototyping and production methods is a key factor in the emergence of new creative tools, it should be noted that this has been made possible by heightened technological complexity and, necessarily, profound changes in the skills required to conceptualise the tools.

In 1700, Bartolomeo Cristofori created the first piano using construction techniques largely borrowed from cabinet-making. In fact, woodworking remained a dominant — and

---

[*] See the interview on pages 66–71.

[†] See the interview on pages 72–78.

[‡] See the interview on pages 59–65.

relevant — craft until at least 1900 when Frank A. Brownell, also a cabinet-maker, designed the *Kodak Brownie*, considered the first camera accessible to everyone. The 20th century brought multiple technological revolutions; Robert Moog, who may have borrowed some cabinet-making techniques for his legendary 1964 Moog synthesiser, drew primarily on his training as an electrical engineer. By the late 1980s, one needed knowledge of neither woodworking nor cable soldering to make a creative tool. The Knoll brothers' development of Photoshop — facilitated by Thomas Knoll's doctoral research in computer vision — exemplified this shift, as the toolmaker's expertise began to migrate from the physical workshop to the computer.

Today, designing creative tools requires a wide palette of skills and a certain technological fluency, including in software development. While designers don't need to be experts in every field to come up with new ideas, most will agree that being able to prototype helps advance one's thinking. In that respect, AI may constitute as great a threat to design as it appears to be an opportunity for designers. Indeed, while, at a surface level, it allows anyone to create software applications without prior coding knowledge, it also establishes a new paradigm in which the underlying principles are never entirely accessible to humans.

If using AI for everything that needs to be designed is our only answer to heightened technological complexity, we risk gradually losing our agency and missing learning opportunities at both individual and disciplinary levels. Ultimately, this may lead to accelerated standardisation of the design of creative tools and, most likely, of what people ultimately create with them.

How Can Design Help?

Ask anyone who's read a book on design thinking what a designer might do differently, and you'll probably hear a variation of 'they would focus on the user' — as if not being a trained designer meant being unaware of the user's existence, or as if the act of focusing were enough. In photography, simply focusing on a subject does not produce a photograph; it produces an image which happens to be in focus.

'the central question that will determine the evolution of our culture is fundamentally one of design: which tools do we want in order to build our creative future?'

Designers can make a difference, but not merely through methodologies — and certainly not by adhering to quasi-religious mantras. At a time when everyone seems to wonder where creativity is headed, and every week brings a new tool powered by generative AI — often driven by what is technically possible — the central question that will determine the evolution of our culture is fundamentally one of design: which tools do we want in order to build our creative future?

As a human activity, creation is like nothing else. It presents a paradox: we create solely for the sake of doing it, yet these acts produce the tangible outcomes that shape our culture. Consequently, the tools we use to create are unique; unlike other designed objects, they do not simply stem from problem-solving. In fact, one of the core roles of the creative tool designer is to ensure that the process does not become centred solely on efficiency and productivity, which would inevitably narrow the range of possible outcomes. When developing such tools, designers often face metaphysical questions that compel them to reflect on their own philosophical beliefs. The distinction between digital and analog technologies is a prime example — a space where our logical minds try to establish quantitative comparisons when, in truth, they represent two almost irreconcilable dimensions of our reality.*

Designing creative tools isn't simply about refining what already exists. It starts well before *wireframing* the user experience and extends beyond *usability* and *aesthetics*. It is an all-encompassing process that necessitates asking fundamental questions, such as 'How could we create differently?'

The design canons of the last 40 years have led us to believe in comfortable design principles that treat users as if they were toddlers. Yet, designing creative tools requires serious interrogation of our assumptions: *Should a tool be simple to use? Is frustration necessarily bad?* Only by doing so can one find interesting and nuanced answers. Topics such as *automation*, *friction*, and *limitations* take on an entirely different meaning when viewed not through the lens of vacuous consumption but through the experience of navigating a creative journey — one that can be exhilarating, torturous, radical, excruciatingly long, and never a simple straight line.

Finally, the designer's lens is one of cross-disciplinary curiosity. Good design draws from a wide breadth of inspiration. While a genuine interest in the creative activity at hand is essential, knowledge of developments in other creative fields is often the source of great ideas. The role of the designer is that of an aggregator, a particle accelerator — one that ensures we never settle for the status quo, or pursue a direction simply because we can without questioning the long-term impact.

* See Analog or Digital? on pages 35-40.

Journey and Destination

When considering the design of a creative tool, one may be tempted to treat it as any other object. Yet, unlike most designed items, a creative tool can never be an end in itself; it exists solely to mediate something greater: human creation. Hence, designing a creative tool is not about offering an answer, but formulating a question: What might it enable one to create?

Beyond Function and Experience

The question of *function* has preoccupied design discussions since at least the time of Louis Henry Sullivan and his famous 'form follows function' dictum.†[26] While knowing how to balance form and function is the minimum prerequisite for a designer to avoid the pitfalls of the ornamental,‡[27] it is not sufficient, on its own, to fully capture the complex nature of a creative tool, for creation cannot be reduced to a function, and it certainly cannot be assessed through usability studies alone.

If we consider the tools used in the visual arts, they all allow the application of colour and shape to a surface. Yet, one could devise all the tests in the world to assess how well they

† This is a common misquotation of the architect Louis Sullivan, who actually wrote: 'form ever follows function'. The addition of 'ever' emphasises the organic and inevitable relationship between an object's purpose and its physical manifestation.

‡ In design, *ornamental* refers to features used for decoration rather than function. While these elements — like motifs or moldings — enhance visual interest, they were famously attacked by Adolf Loos in his 1908 essay, 'Ornament and Crime'. Loos viewed decoration as a waste of labour and a sign of cultural immaturity. This critique paved the way for the Modernist 'less is more' movement, which traded 'extra' flourishes for clean, functional surfaces.

'designing a creative tool is not about offering an answer, but formulating a question: What might it enable one to create?'

perform their *image-making function*, and it would still be impossible, from this perspective alone, to explain why some of them — like oil-on-canvas or, centuries later, Photoshop — ended up transforming creation entirely.

The shortcomings of the form-function duality have long been pointed out by some designers who claim that we have now moved from designing *products* to designing *experiences*. This is certainly true for service companies such as Spotify, Netflix, Uber, and Airbnb, where the product, usually a smartphone application, is only a vehicle to the experience — which is the value customers are really paying for. This is also true for companies that operate at the frontier between the physical and the digital — typically, smartphone manufacturers, where the experience often plays a determining role in how customers choose one product over another, even if their *look* and *function* are nearly identical.

Yet, while considering the *experience* offers greater depth than focusing on *function* alone, creation cannot be reduced to the mere *experience of creativity*. Moreover, history features countless examples of great creative tools where some aspects of the experience are objectively terrible. From the daunting learning curve of the piano to the back pain associated with the violin, or the physical strain of the trumpet, musical instruments offer numerous examples that bring necessary nuance to the role of the experience in the success of a creative tool.

This is not to say that *function* and *experience* are never important for creative tools. In their interviews, Roger Linn and Mats Herding Solberg emphasise the crucial role of the former in tools — something designers too often forget[*] — while Jonathan Yeo points out that a fun experience can be the reason we return to certain tools and eventually progress with them.[†] This holds true even if the goal of creation itself is not intended to be entertaining, as reminded to us by Natalia Ilyin.[‡] However, function and experience are insufficient; when it comes to creative tools, designers need to adopt a broader perspective, one that considers what the creator is trying to achieve and the journey that will take them there.

Result-Oriented Creation

'Good artists copy; great artists steal.'

— Steve Jobs, 1996

Whether this quote is about painters or poets, or whether it comes from Picasso, T. S. Eliot, or Steve Jobs, nobody seems to agree.[28] Yet one thing is clear: it serves as a reminder that creation often begins with *imitation*.

Indeed, those who wish to create something may choose to use the same tools as those 'sitting next to them',[29] or work toward a result they can observe — a tradition spanning from the apprentices of the classical world to the buzzing bottegas of the Renaissance, and the YouTube tutorials of today. In the learning phase, the result one can hope to achieve with a given tool takes over all other considerations that may have preoccupied the designer. Aspiring artists may put up with months of learning complex 3D modelling software cluttered with menus, while others may painfully adapt to musical instruments designed to achieve a certain purity of sound at the expense of ergonomics.

---

[*] See page 53 and pages 60–62.

[†] See pages 68–69.

[‡] See page 77.

Interestingly, veteran creators may be similarly goal-oriented and will likely invest time and money in new tools only if these allow them to achieve something specific — not if they look good or provide a great experience. This is why creative tool manufacturers often try to attract customers through the introduction of new features framed around the results they can allow for achieving: cameras with higher shooting rates *to ensure the decisive moment is not missed*, faster 3D engines *for more realistic renders*, and portable music instruments *to capture better musical ideas, when they occur*.

In that sense, creators often don't really care if a tool is perfectly functional or provides a great experience, as long as they *believe* it will allow them to achieve a certain result. However, this certainly does not mean that all creators have the same outcome in mind when they look at a tool. Therefore, the danger for the designer is that, by trying to accommodate everyone, they may adopt a feature-focused approach — risking reducing design to the mere assembly of a technical specification and ignoring another crucial aspect of creation: the *journey*.

It's Not the Destination, It's the Journey

In his 1911 poem 'Ithaka', Greek poet Constantine P. Cavafy proposed a great metaphor for life: Wherever we think we are going, the whole point is not that we get there eventually, but how we do.[30] As it happens, it is a great metaphor for creation too — one that may be more approachable than speaking of *autotelic* activities, as Mihály Csíkszentmihályi does:[*] Indeed, it reminds us that even if creators can be result-oriented when they choose a creative tool, the creative journey is paramount in the act of creation.

From the designer's standpoint, focusing on the creative *journey* allows for accounting for two crucial elements of creativity: *time* and *agency*. Where, typically, function and experience are assessed globally — an experience is either good or it is not, and a function is either achieved or it is not — the *journey* allows designers to escape the trap of trying to average everything into the *satisfactory*. Instead, they can embrace the fact that different stages of the creative journey can involve entirely different experiences: *fun* during onboarding, *challenge* during learning, and *reward* at a later stage.

Of course, popular design frameworks and tools, such as *user flows*[†] and *wireframing*, also consider time. Yet they usually do so to ensure that all potential scenarios are accounted for and that users are not lost in the labyrinth of choices they face. In many respects, a creative tool seeks precisely the opposite: the creative journey belongs to the user — not to the tool, and certainly not to its designer. In fact, if it comes to that, losing oneself is often part of creating something meaningful.

Finally, putting emphasis on the *journey* reminds us that creation does not always culminate in a tangible result such as an artwork or a sculpture. A performance, for instance, is also a form of creation, even if it often involves reproducing a pre-existing work: for instance, a sonata, a play, or a DJ set. Some performative arts rely extensively on *improvisation* — the creation of something essentially ephemeral. In such cases, tool design can play a determining role in the evolution of a discipline by enabling greater improvisation in performances.

Modern DJing tools, for one, allow for increasingly complex manipulations of music in real-time. As a consequence, DJs can now

[*] See page 16.

[†] Sometimes called an *experience map* or *user journey* — though the latter should not be confused with the *creative* journey discussed here.

operate at the intersection of the *set* — which traditionally focused on curating a playlist — and *live* music — layering multiple musical elements, some from commercial tracks, some from their own production, and some improvised on the spot.

When it comes to creative tools, designers must transcend the traditional dualism of *form* and *function* and push beyond the comforting precepts of good design* — otherwise, they risk building an unobtrusive product that sits on a shelf only to spark the occasional conversation. Similarly, if a great *experience* becomes the designer's primary goal, the risk is ending up with a toy or a device for consumption, not a tool for *creation*. Designing a creative tool is a delicate balancing act: It must provide the technical power to reach a destination without stripping the creator of their agency or the transformative wanderings that define the creative journey.

'Designing a creative tool is a delicate balancing act: It must provide the technical power to reach a destination . . . [and] the transformative wanderings that define the creative journey.'

* A framework by German industrial designer Dieter Rams that focuses on ten rules or attributes of the product. It states that design must be: innovative, useful, aesthetic, understandable, unobtrusive, honest, long-lasting, thorough down to the last detail, environmentally friendly, and involve as little design as possible.

On Friction & Limitations

It is an understatement to say that friction has had a bad reputation in design: it slows things down and, quite likely, generates frustration. As for limiting the very tools that are supposed to extend our creativity, many fear it will limit what can be done with them. But what if friction, by slowing down time, were the secret ingredient that allowed for a richer creative process? What if frustration was not an issue after all, and limitations were actually opening up new creative perspectives?

When Friction Is Desirable

At present, these questions are often ignored because it has become a universal design principle that friction is the enemy of the user, and hence of the designer, too. The reasoning is that friction distracts users from the goal they are *really* trying to achieve by consuming precious cognitive resources on futile tasks — tasks that could have been avoided had they been identified during the design process. And indeed, most people will have experienced the nerve-racking feeling of navigating inane menus when trying to operate a Smart TV or pair a smartphone with a car's audio system.

However, as sensible as it may sound, this view is skewed by the fact that design discussions revolve largely around tightly framed scenarios in which users are trying to achieve specific, and often isolated, actions — for instance, navigating a website or consuming content. Moreover, the primary reason removing friction has become a quasi-religious design mantra is that, in such scenarios, the fluidification of interactions typically results in more monetary transactions.

Yet, design is not solely about consumption or efficiency. Creative tools don't *need* quick user interactions and, in fact, in some cases, friction

'friction, by slowing down time, is the secret ingredient . . . for a richer creative process.'

is *precisely* what makes for desirable tools. Modular synthesisers, for instance, require users to assemble their own instruments using modules from various manufacturers, each with its own speciality. Once they have a rig, they need to patch it — that is, connect the different modules with patch cables. Finally, each button, slider, and knob — of each module — needs to be adjusted to get something out of it all. This long and apparently tedious process is precisely why some musicians prefer modular synthesisers; they regard it not as a futile task that could have been avoided, but as the very essence of their creative journey.

In a 2016 paper, researchers at UCL in London argued that friction can have positive effects and create more mindful interactions, for instance by guiding 'the user towards a particular course of action without having to rely on willpower alone.'[31] In other words, it is precisely by providing creators with something predictable to focus their minds on that friction avoids depleting precious cognitive resources — which can then be allocated to creation when inspiration comes. While no one likes an absurdly frustrating user experience in which nothing makes sense, well-dosed friction can be part of the designer's repertoire to enhance the creative potential of a tool.

If the term *friction* has been largely used metaphorically so far, it is interesting to note that our feelings toward actual — physical — friction are generally positive when it comes to interfaces. In fact, many people prefer older interfaces *because* of friction: the nice feeling of turning knobs on a vintage amplifier comes from friction; the same applies to older camera lenses, especially those made of metal, where friction provides the right amount of resistance to ensure smooth focusing.

And one cannot fail to mention the feeling of pen on paper: friction is not only what makes us feel the lines as we trace them; in *Creative and Mental Growth*, American professors Viktor Lowenfeld and W. Lambert Brittain proposed that the enjoyment of kinaesthetic* sensations while scribbling in childhood is the very mechanism by which we discover the connection between our physical motions and the first marks on paper that allow us to express ourselves as humans.[32]

In physics, one effect of friction is to make movement take more time — or energy. The creative journey is a movement that takes place within the mind of the artist. By pacing the process, friction not only induces greater engagement but also enhances the sensory experience of creation by making the artist quite literally feel the journey. Along the way, it increases the chances of unexpected discovery and fosters a deeper sense of ownership.

What If Friction Leads To Frustration?

At the turn of the 1980s, the introduction of the Graphical User Interface† enabled computers to become personal devices that could be operated by non-experts. Today, one only has to look at the relationship people have with their smartphones to measure the success of something that started as an experiment in a

---

*Kinaesthetic: Relating to kinaesthesia — the sensory perception of movement and the physical position of the body's parts.

†Often just GUI, or UI.

research lab.* Although it may sound counterintuitive, what the GUI has brought to the table can, in many ways, be summarised in one word: *friction*. By moving interactions from typing command-line instructions to visual choices activated by clicking icons, it made computer operation more accessible — but also less open-ended. This largely explains why, to this day, software developers and IT professionals often bypass GUIs altogether and operate their machines via the command line.

Some users are not interested in the 'positive effects' of friction if it results in inefficiencies. For the designer of creative tools, it may pose a dilemma: how to use friction without generating frustration? One way to answer this question is to consider a more nuanced definition of frustration — one that examines the user, not at a single moment in time, but over the course of their *journey* with a tool.

Compared with most other toys, a LEGO set does not allow for immediate play. Instead, it needs to be assembled, brick by brick. Soon, the initial frustration of the child, whose motivation was to play with a toy, gives way to a richer experience and a stronger reward — often referred to as the 'IKEA effect'† — and endless creative possibilities.

When designing the OP-1, the team at Teenage Engineering made a curious choice: they did not implement an undo function in recording mode. While it would have been technically possible, they decided to stick to the metaphor of the analog tape recorder and the restrictions that come with it. It did not hinder the success of their product — far from it — and while it remains impossible to be in the heads of all OP-1 users, it is fair to assume that they have learned to work with and find inspiration in these restrictions, not determined by technology, but by design. In fact, this concept has remained consistent in subsequent successful designs from Teenage Engineering. What may have initially felt like frustration undoubtedly encouraged a raised sense of awareness and a more deliberate approach.

This highlights an important point: *what is deemed frustrating is largely subjective, shaped by individual abilities, acquired skills, experiences, and goals*. In our earlier computer interface example, it is worth noting that even the most advanced users once had to navigate a learning phase where command lines were anything but natural and likely quite frustrating. Even when it causes frustration, friction can be a source of learning, provided it affords the user space to overcome the challenge and ultimately become proficient, as new neural connections are developed in the process.‡[33]

(Un)Limited Tools

Regarding functionality, there is a common perception that a limited tool yields limited creative output. SketchUp, for instance, the

* In the 1970s, researchers at Xerox PARC pioneered the graphical user interface (GUI) and the mouse, shifting computing away from complex text commands towards visual interaction. While Xerox did not fully commercialise these ideas, the 'experiment' provided the fundamental blueprint for personal computing and, ultimately, the intuitive, touch-driven interfaces that define the modern smartphone experience.

† A cognitive bias where people place a higher value on objects they've partially built or created themselves. In the LEGO context, overcoming the initial frustration of assembly fosters a deeper pride and attachment than a pre-built model.

‡ This is supported by the seminal research of Frey and Morris (1997), published in Nature. Cited over 2,000 times according to Google Scholar, this paper posits that the brain marks new experiences with a 'synaptic tag', but only a strong, effortful challenge triggers the production of the 'protein glue' required to lock that tag into a lasting neural connection. Without this struggle, the brain lacks the signal necessary to convert a fleeting experience into long-term proficiency.

affordable 3D CAD* software widely used in architecture, has a simple set of tools that produce images with a distinct hand-drawn look some architects find characteristic of amateur work — not without a hint of snobbery. However, having a *fingerprint* does not necessarily limit the creativity of output. Many would argue that the strength of a tool like SketchUp is its ability to enable rapid design and visualisation of a building without having to learn endless menus or becoming a CAD software expert.

If limited tools constrain, unlimited tools, by contrast, are at risk of *bloating* as new functionalities appear. So much so that designers need to develop strategies to accommodate different user types: Adobe has long struggled with the dual issue of veteran users who don't want to lose their favourite functionalities, and emerging younger generations who have little interest in learning legacy ways of doing things. As a result, Adobe often proposes two or more distinct interfaces and, in some cases, maintains two versions of their tools in parallel — one on desktop, usually more advanced, and one web-based, simplified and built for collaboration.[34]

But as much as contextual or disappearing interfaces can appear to reduce the perceived complexity of a tool, the *existence* of choice remains an issue. As music-making tools become richer in sounds and features, many musicians experience *choice paralysis*, spending hours auditioning thousands of sound presets rather than making music. One way to solve that issue is to leave choice to the tool by automating as much as possible — which, ironically, brings another, more insidious, form of limitation.†

'Limitations serve as the frame for a creative tool . . . carving out the contours of what a tool can achieve.'

In *The Interface and the Potential*, Australian design researcher Stephen Farrugia argues that the problem doesn't have to do with interface design per se, but rather with 'the potential for distraction that lies behind the interface.'[35] Taking the example of the computer, he explains: 'When we're using a computer, the interface isn't the centre of attention as much as the potential of the computer is. When you are aware of it, potential is attractive and powerful. Computing potential is powerful in how generally useful it is. General usefulness is harder to ignore, and its absence is easier to notice.'[36] In other words, the more a tool can do, the more inherently distracting it is.

Hence, it is important for the designer to always allow themselves to step back and consider their options: sometimes it is about embracing technological limitations; sometimes it is about making a deliberate choice to limit the tool in order for it to remain true to its promise.

The reMarkable tablet is a great example of a tool that fully embraces the limitations of its underlying technology. Unlike fast colour displays, e-ink displays are unsuitable for watching movies, let alone playing games, as they exhibit slower response times and image ghosting. On the other hand, they are highly suited to additive processes, such as sketching or writing, and prolonged viewing. As a consequence, reMarkable products exhibit a natural alignment between technology and design intent, thereby encouraging deep user focus and the generation of better ideas. Moreover, its designers maintain purposeful limitations to ensure this always remains true. In his interview, Mats Herding Solberg, who has

* Computer-Aided Design (CAD) software is used to create, modify, analyse, and document 2D or 3D graphical representations of physical objects.

† See The Temptation of Automation on pages 31-35.

been leading the design at reMarkable since the early days, recounts how keeping their tablet limited has remained one of the hardest things, but also the key to being able to guarantee 'better thinking' to their user base.*

'The typewriter unlocked the rest of it for him. At first it was slow, so slow, and the constant typos were frustrating beyond belief. It was as if the machine was actively — but slyly — opposing his will.

But when he got better at it, he began to understand what the machine really was — a kind of magic conduit between his brain and the blank page he strove to conquer.'

— Stephen King, 1990[37]

Human history is teeming with artists who embrace the limitations of their tools, from 19th-century impressionists, who developed styles that allowed them to quickly capture outdoor scenes, to modern video game designers, who regularly develop simplified art directions that allow their games to run on slower machines. Most painters and photographers would agree that framing and composition are often the most fundamental elements of their work.

Rather than viewing friction and limitations as negative attributes that stifle creativity, it is essential for designers to consider the focus they provide. Limitations serve as the *frame* for a creative tool: just as framing defines the boundaries of the world depicted in an artwork, limitations carve out the contours of what a tool can achieve. Following the same metaphor, *friction* is the compositional element of the creative tool: in a painting or a photograph, composition guides the viewer's eye; in a tool, friction guides the creative journey.

* See page 61.

The Temptation of Automation

Automation is so common in modern creative tools that its role often remains invisible. While it can democratise creativity by lowering technical barriers and enabling new levels of complexity, it is not a universal good. If misapplied, it risks flattening human expression and erasing the idiosyncrasies that make creative works unique. For the designer, the challenge lies in strategically directing automation: simplifying while preserving the human agency essential to the creative act.

What Is Automation?

Most of the creative tools we use feature some form of automation. In fact, it has become so synonymous with the very idea of the *tool* that most people overlook the role it plays in the background. For instance, in the fraction of a second between pressing the shutter and a picture being taken, a camera's autofocus measures the distance to the subject and instructs the lens to focus in order to ensure sharp details. In software tools, the scope of automation is even less tangible. On a daily basis, most of us use a word processor that offers to 'autocorrect' our mistakes and 'autocomplete' our sentences.

Before we can analyse the design implications of automation, we need to consider how it came to be. The design of creative tools has not evolved in a vacuum, but rather as part of a larger movement encompassing the design of everything from kitchenware to the built environment. From that perspective, one cannot comprehend our relationship to automation without examining the influence of the Industrial

Revolution[*] on design at large — an influence that is still felt today. At the core of the acceleration of production, and thus the reduction of costs, was the idea of transforming energy to power machines which could then replace manual labour, especially in tasks judged repetitive or inefficient. Although other forms of automation, such as clockwork, preceded automated manufacturing, the Industrial Revolution marked a turning point and, to a large extent, the birth of modern automation.

The effects it had on humanity's creative output did not go unnoticed; in many respects, the Arts & Crafts movement arose in reaction to the perceived qualitative degradation of design resulting from the automation of production, which standardised centuries of vernacular tradition. Often considered one of the most influential design philosophies of modern history, it reshaped how we think about design in the context of mass manufacturing and raised questions about the place left for the craftsperson.

It did not halt the broader movement toward the automation of all things, and today the question of automation has become significantly more complex. Indeed, while traditional tools and machines were, for a long time, clearly distinguishable — the latter usually being much larger in size — most modern tools have effectively become machines themselves, some even fitting in the palm of a hand. In effect, miniaturisation and digitisation have made the boundary between the human-driven and the automated largely invisible.

[*] A period of significant socio-economic and technological change, beginning in Great Britain in the mid-18th century. It marked the transition from hand-production methods to machines, the increased use of steam and water power, and the development of machine tools. This shift fundamentally altered the way artefacts were produced and paved the way for modern manufacturing.

In addition, whereas during the Arts & Crafts movement automation impacted mostly manufactured goods and their design, it now touches almost all forms of human expression. Yet, while the need to create faster or at lower cost may be appealing in areas such as VFX,[†] it is not a universal rule that everything should always be automated. Hence, what can be expected from the designer of creative tools is, at the very minimum, to have a clear idea of what they seek to offer through automation.

What Automation Can Do

The first and obvious benefit of automation is *simplification*. While simplicity is not necessarily a required quality for a creative tool, there are specific contexts in which it can make a difference. Skill-building is one such context, and automation often helps smooth the learning curve. In the camera example above, autofocus — and autoexposure — have helped generations of photographers take their first pictures without having to worry about technical parameters, allowing them to concentrate on the more fundamental aspects of photography: framing and composition.

By doing so, it enabled them to experience the rewarding feeling we get when we create something we like. For some, it may have been just what they needed to feel empowered to continue until they could truly benefit from the added agency of knowing how to use a camera without any automation. In other words, simplification made the tool more *accessible*.

But accessibility is not just a question of simplicity. It is also about the diversity of users to whom a tool may *speak*. In some cases, automation can open a field of creation that was previously restricted to people with certain talents and abilities. In *Multiple Intelligences*,

[†] Visual effects such as the incorporation of virtual 3D elements in films and TV series.

American psychologist Howard Gardner suggests that we possess not one but multiple intelligences, which largely determine our natural talents and abilities in life.[38*39] He notes that, while we all have different intelligence profiles, we often mistakenly assume that others think like us.[40] He uses the example of school exams that end up measuring not how well the person has understood the material, but how much their mind resembles that of the person who designed the test.[41]

Prior to the introduction of the Roland TR-808 in 1978, playing the drums required excellent physical coordination and, admittedly, a good ear for music. With step sequencing — a form of automation allowing the programming of drum loops using a grid — a new generation of musicians began composing rhythms. They did not replace real drummers where real drumming mattered, but they did create new musical genres such as techno and house. Automation enabled rhythmic composition to move from being reserved exclusively for those with high bodily-kinaesthetic and musical intelligences to being accessible to those with more visual and spatial forms of intelligence.

Finally, there are times when automation enables us to create things that were simply not possible otherwise — or at a level of complexity that most humans are unable to handle. Parametric tools such as Grasshopper for Rhino 3D, a modelling software favoured by some architects, have led a mini-revolution in the design of buildings, evident in the heart of our cities. With it, designers can automate their processes — for instance, to generate thousands of wall-panel variations that, once manufactured and assembled, follow large-scale

* Gardner first introduced this framework in his 1983 seminal work, *Frames of Mind: The Theory of Multiple Intelligences*, where he proposed that human cognitive ability is composed of distinct, autonomous modalities.

▼ *Heydar Aliyev Center in Baku, Azerbaijan designed by Zaha Hadid Architects (2012)*

organic curves, as seen at the Heydar Aliyev Centre in Baku, Azerbaijan.

'automation risks flattening the differences in the creative outputs of humanity as a whole.'

Automation Considered Harmful

It would be wrong to think that automation is always a net positive. If it makes creation too effortless to be truly engaging, it can derail the reward loop and turn the process into a bland activity. In contrast, building something *integrally* triggers positive reinforcement, as the brain interprets the finished work as a direct, rewarding reflection of every *choice made* and every *action taken*. The most vocal critics of automation emphasise that it is precisely the repetition of operations, especially manual ones, that enables us to build the neural connections necessary for truly acquiring a skill.

In a 2013 article for *The Atlantic*, American writer and journalist Nicholas Carr explains the 'generation effect' — a phenomenon observed by psychologists since the 1970s: 'When you repeat the same task over a long period, your brain constructs specialised neural circuits dedicated to the activity.' He contrasts this with automation, which 'turns us from actors into observers. [...] That shift may make our lives easier, but it can also inhibit the development of expertise.'[42] In other words, tools that don't leave space for a healthy dose of repetition might not allow for learning other than superficially.

Moreover, if we have so far limited the scope of the discussion to the perspective of the creative individual, the greatest danger posed by automation is undoubtedly on a much larger scale. By simplifying, adapting, and making tools more accessible, automation risks flattening differences in the creative outputs of humanity as a *whole*. If one were to blindly compare two piano players — one a beginner and one a professional — it would immediately be evident which is which upon hearing them play. A similar experiment with a modern instrument, such as a modular synthesiser, may not be as conclusive — even if 'aesthetic fluency'* can explain these results in part. By correcting mistakes, smoothing lines, and aligning elements to grids, automation is already affecting *happy accidents* and the subtle nuances that make what we create unique, and this effect is likely here to stay.

Where To Direct It

From a design standpoint, automation should be considered only for tasks that offer little potential for creative expression. For instance, while one could argue that there are infinitely many possibilities for creative experimentation with a camera lens's focus, in practice, most photographers would agree on what constitutes *good* focus. Hence, if autofocus gives them the freedom to direct their attention to other aspects of photography where they feel they can have greater creative control, they will gladly embrace it — especially if they can modulate its behaviour to their taste and deactivate it when needed.

The idea that the time or effort saved by automation can be redirected toward more engaging aspects of the creative activity is crucial. DJs, for instance — who largely moved

* The 'knowledge base' or 'vocabulary' an individual acquires through exposure and education to process art. While most listeners possess a high cultural fluency for piano performance, they may lack the specialised literacy required to discern technical mastery within electronic music interfaces.

away from manual beat-matching* once equipment began to handle the task for them — are now much more invested in layering multiple tracks and using complex juxtapositions of loops. This would simply not have been possible had they still been spending half their time adjusting the tempo. While critics denounce the modern DJ performance as symptomatic of a society that favours style over substance, it seems that modern DJ tools allow for as much depth of expression, if not more, than those in use three decades ago. Similarly, automated coding allows designers to prototype quickly while focusing on the core experiment rather than getting stalled in learning how to code or, even earlier, in the tedious process of decomposing a vision into a technical brief.

As we witness its effects across all creative fields, yesterday's great temptation of automation is giving way to a new paradigm in which implementing automated formulas has become paradoxically simpler than designing the interfaces that could support meaningful interaction and adequate creative control. In that context, the emergence of new AI-driven tools will likely require even greater vigilance from designers seeking to develop tools that avoid reducing creation to a formulaic form of expression.

* A foundational DJ technique where the playback speeds of two records are synchronised to ensure their beats overlap seamlessly. In the analog era, this was achieved by ear using a pitch fader; on most modern digital DJ decks, this process is automated through software, often referred to as a 'sync' function.

Analog or Digital?

Of all the recurring themes in discussions about creative tools, the question of whether analog tools are better or worse than their digital counterparts often causes confusion. This is largely because the conversation is constantly at risk of falling into one of two traps: the technological comparison, which is too theoretical to be useful to most people, or the *collector's debate*, which focuses more on the prestige of certain items than on what they actually allow creators to achieve. Moreover, the latter tends to further confuse the issue by bringing in aspects of interface design, material choices, or quality standards — elements that are all important to a design discussion, but are not strictly linked to the difference between analog and digital.†

Yet, there is value in understanding what *analog* and *digital* refer to, and what each is respectively good at, especially for those who design creative tools. This is why this section seeks to establish a current state of play by looking at the specific properties of both digital and analog tools through a design lens focused on universal rules — that is, rules applicable to all categories of creative tools.

What Do 'Analog' and 'Digital' Mean?

We refer to tools as *analog* when they convert

† The impression that analog tools possess superior quality often stems from emergent properties — especially in interfaces which inherit characteristics dictated by their underlying components and physical manufacturing. These are frequently compared to the poor design choices of modern digital interfaces, which are not inherent to the medium itself. If rigorous design is applied, interfaces for both digital and analog tools can be equally high-quality and effective to use.

an input into an output through continuous physical changes. In the Palaeolithic flute, one of the first known creative tools (ca. 60,000 BCE), sound arose from the vibration of air forced through holes carved into the bone. This principle of physical conversion defined most tools until the mid-20th century. For instance, a traditional piano converts key strikes into string vibrations, a film camera chemically captures light on celluloid, and a painter's brush translates the continuous pressure of the hand into a physical trace of pigment on the canvas. Similarly, the needle of a DJ turntable generates sound by moving inside the vinyl record's groove. Even early synthesisers, though vastly more complex than the bone flute, rely on a continuous physical change — replacing air vibrations with fluctuating electrical voltage.

The historical pivot arrived with the integrated circuit and the computer. Instead of maintaining a physical *analogy*, the computer chip enabled the ability to encode any input as a discrete stream of binary bits (0s and 1s) — *digital* information. This marked a shift from the physical to the abstract: the digital tool does not function through physical *conversion* but rather through a *description*.

A description of something is never going to be the thing itself. For instance, if someone describes what they saw, thought, or heard, no one else will ever be able to see, think, or hear it as they did. Yet, if there is no limit to how detailed a description can be, it might capture so much of the reality being described that we cannot, in effect, distinguish between them.

A digital piano measures key velocity to calculate or select a prerecorded sound sample that best represents the loudness of the note played; a digital camera detects light using a sensor constructed as a pixel grid, assigning numerical values for colour and intensity at discrete positions; and a sketching app converts stylus movement and pressure into coordinates and data values to render a drawn line. Ultimately, all these inputs are reduced to discontinuous binary code formatted the exact same way, regardless of whether they represent a line, a photograph, or a symphony. Understandably, these nuances are often lost on the general public. Whether analog or digital, all these tools allow for creating music, images, objects, and art in general! Furthermore, from the outside, it is hard to tell them apart — which makes it even more difficult for designers, who are more accustomed to focusing on how things look, feel, and function than on their inner workings. Yet the distinction has profound consequences for how we use these tools and, ultimately, for what we create with them.

Which Is More Authentic?

When justifying their preference for analog tools, artists often invoke *authenticity*, and specifically authenticity of the output. Indeed, if one considers cameras, most photographs from the period when film was dominant display natural-looking colours; in fact, the very idea of systematically editing and retouching images really took off with digital workflows. Similarly, an analog synthesiser produces original sounds directly, using electrical signals, rather than by *emulating* them, as a software plug-in would.

But these comparisons examine only one side of the story. Before the advent of colour film, photographs did not exactly provide a true representation of reality, for they were all in black and white. Moreover, while often neutral-looking — though not always — colour films are calibrated by manufacturers to match a specific style. In comparison, a digital camera can produce a technically neutral image, clearly separating capture of light levels in the scene from choices about the look-development of the final picture — leaving more agency to the photographer. While editing and image manipulation have become easier with digital tools, they have long been practised through

> *‘analog tools impose uniqueness and character . . . Digital tools remove uncertainty and allow for multiple exact reproductions — which requires us to reconsider how we value the output of a creative process.’*

techniques such as optically layered composites, double exposure, or even painting directly on the film negative.

As for analog synthesisers, their very name reminds us that their *raison d’être* was never to sound authentic but, indeed, *synthetic*. In fact, the introduction of digital drum machines was partly in reaction to the analog ones that sounded artificial. By using individually recorded digital samples of real acoustic drum sounds, the Linn LM1 provided musicians with both the convenience of a programmable drum machine and the sound of *authentic* drums.

If authenticity is an interesting question in itself, it does not necessarily provide a conclusive angle for comparing analog and digital tools. It does, however, raise the important question of the *uniqueness* of the output. A traditional painting, for instance, only exists in one unique embodiment — it cannot be reproduced exactly, ever. In analog photography, each print from a negative is slightly different from the next, depending on chemical variability and the process of the person developing it. Hence, with uniqueness comes *character*. With analog synthesisers, as with traditional physical instruments, each copy — even when it comes from the same factory — has its unique characteristic acoustic signature.[*] And it won’t sound the same in a month, in a year, or in ten years’ time.

Conversely, digital tools enable the creation of exact replicas of any artefact digitally produced, whether a photograph, a digital painting, or a synthesised symphony. Analog-like character can be simulated, but by default, digital tools produce perfectly repeatable results. Typically, a digital synthesiser, whether physical or software-based, has *presets* that allow saving and recalling of an exact sound with a single click.

In short, analog tools impose uniqueness and character, which can be limiting but aligns philosophically with what many artists seek to achieve. Digital tools, on the other hand, break away from these limitations by removing uncertainty and allowing for multiple exact reproductions — which requires us to reconsider how we value the output of a creative process.

Which Is Easier?

Ease of learning and ease of use, supposedly leading to being more prolific as an artist, are often cited among the main reasons to choose digital tools over analog ones. In industrial design, a single 3D model enables the generation of multiple images of a product from different angles and lighting conditions, and the evaluation of different colours and materials for each one. Digital turntables such as the Pioneer CDJ can automatically beat-match two tracks — that is, make them play at the same tempo — whereas it would take months for a DJ to learn and master the same technique using analog turntables.

However, it is not only new processes or intelligent features that matter; the very fact that digital tools have few limitations when it comes to resources plays a role. For instance, many would argue that the ability to take 10,000

[*] In March 2025, the ‘Baron Knoop’ Stradivarius — a 1715 model — sold for a record-breaking $23 million, making it the most expensive violin ever sold. This highlights how physical ageing, provenance, and manufacturing variations create a finite, non-reproducible resource.

'By having natural limitations, analog tools encourage intentionality and commitment . . . Digital tools offer unlimited freedom.'

digital photographs on a single memory card makes it easier to capture a good one. Yet, the opposite of all that is also true! If it is just to communicate a quick idea, a paper sketch is likely much easier to make than having to model something in 3D — and if sketching seems a hard skill to hone, learning 3D modelling certainly takes time too. As for beat-matching, while digital turntables have literally made it as easy as the press of a button, audiences in the age of digital music expect more from a DJ than just clean track transitions. Today, preparing a DJ set often involves spending hours on a laptop, slicing tracks, and placing cue points to perform complex looping and layering live. Finally, if taking more photographs increases the likelihood of finding a good one, locating it in a folder full of nearly identical images is anything but easy. So much so that AI-powered curation is now a feature of photography applications such as Lightroom.

What some might consider easy, others might consider hard. And what makes some more prolific slows down others. The reason for these differing perspectives is not merely a question of subjectivity but rather reflects real objective differences. Analog tools often rely on tangible resources, which limit their use. Typically, completing a painting on canvas requires sufficient paint — and probably more than one working canvas. Similarly, analog photography relies on the use of film, and analog turntables require the DJ to bring along a heavy crate of vinyl records. In addition, most analog tools are dedicated to a specific use. As a consequence, acquiring new tools often involves disposing of older ones — or relocating to a larger studio.

Digital tools, by contrast, often have very few limitations, whether in the resources they use* or in their physical embodiment, and — in the case of software — are readily available shortly after a download request. Famously, any action can be undone, and, in software form, multiple digital tools that replace an entire studio can fit inside a laptop that can be carried and used anywhere.

By having natural limitations, analog tools encourage intentionality and commitment, which often aligns with their dedicated interface, enabling a deeper kinaesthetic† bond and increased focus — without notifications or surprise mandatory updates. Digital tools, in contrast, are more forgiving, more flexible, and offer unlimited freedom.

Which Has the Best Quality?

If someone is going to spend a considerable amount of time creating something, they usually want to make sure the quality of the result is going to be worth the effort; questionable pigment choices have been responsible for the degradation of some of the world's most renowned paintings. Therefore, it is not surprising that, when it comes to choosing between analog and digital tools, the question of quality is often central. It has become a common belief that, if it has not already, digital will probably win this race at some point, thanks to seemingly endless technological progress in computer chips — which are now part of every digital tool. The steady increase in digital camera and digital audio resolution, and the infinitely zoomable vector lines and shapes of

* This refers to *resources* as the user perceives them. In reality, digital tools depend on electricity, cloud storage, and — increasingly with the rise of AI — cloud computing. All of these ultimately rely on finite resources and have a significant environmental impact.

† The sensory perception of movement and touch.

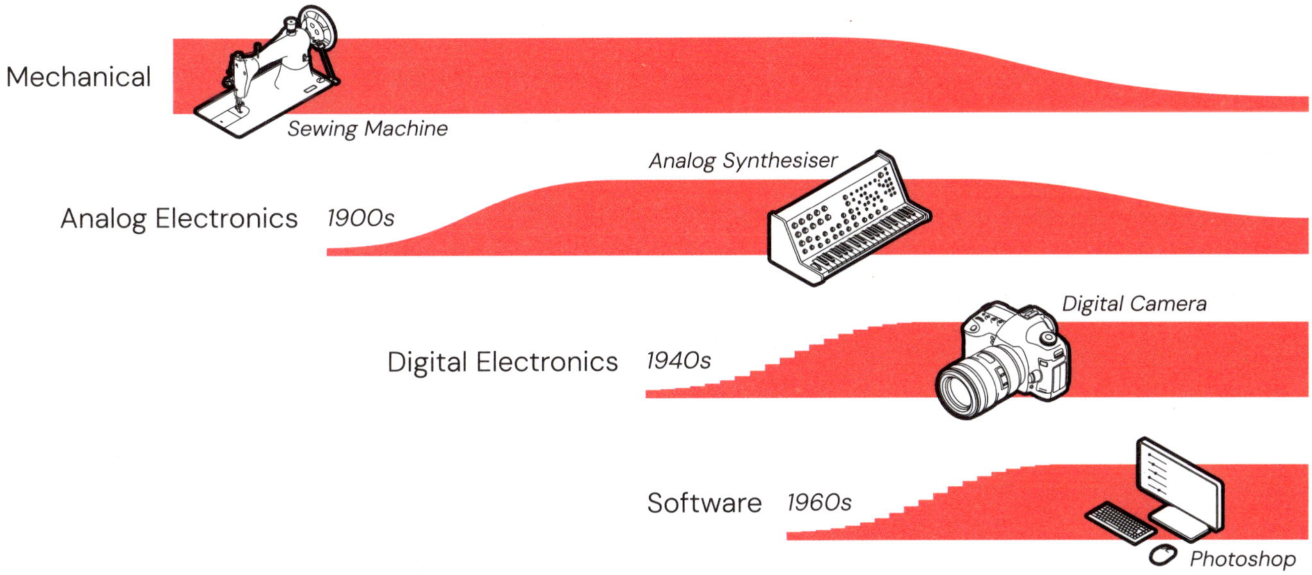

Analog	Digital
Continuous physical change.	**Discontinuous** abstract description.
Infinite detail with no hard-stop, down to the atomic scale.	**Limited** to detail captured in the description — missing information has to be guessed.
Unique and never exactly reproduced.	**Reproduced** exactly, multiple times.
Imperfect and has individual character. **Evolves** with time or situation — such as age or temperature.	**Precise and certain** — individual character and evolution have to be simulated.
Limited resources — such as availability or size — can encourage **focus** and **commitment**. **Dedicated** physical interface can enable deep kinaesthetic interactions.	**Flexible** resources can be immediate and forgiving — unlimited freedom (at the risk of excess, distraction and interruption).
Material existence can embody human experience, interaction and meaning.	**Virtual** existence trades materiality with possibility and perfection.

digital illustrations, all seem to go in this direction.

However, those are just driven by numbers — numbers that happen to compare distinct properties. An analog camera's film has no concept of resolution, per se. Sharpness may degrade as one looks closer, but grain, unlike a pixel grid, is not a hard stop. Similarly, when a musician plays an analog instrument live, whether acoustic or electric, there is no notion of *sampling*, as sound waves fly directly into the ear of the listener with an infinite resolution — not unlike the infinite resolution of a hand-drawn illustration on paper. In fact, associating *resolution* and *infinity* is both a physical and conceptual oxymoron.

This is not to say that numbers never matter: they are indeed useful for comparing two tools, provided that both are digital! For instance, the fact that the latest digital camera has more pixels than any from the 2000s is a strong indicator of a comparatively higher quality image. However, irrespective of the number we pick, quality itself may not be that useful a consideration when it comes to making general comparisons between digital tools and analog ones; for, under the right circumstances, both are perfectly capable of delivering results of high enough quality for most artists' needs.

How they achieve that raises a more interesting question: Historically, analog tools have always captured the entirety of the human action — up to the atomic arrangement of paint particles on canvas in the case of painting. Conversely, digital tools describe everything abstractly using a finite number of selected data points. The more points, the better the description. If more points are needed — for instance, to obtain finer details — interpolation between existing points is often used — sometimes secretly — to fill in the gap. In other words, missing information is guessed.

As technology progresses and AI's role becomes increasingly important in making interpolated data indistinguishable from real physical details, the question becomes increasingly philosophical: analog tools offer perfect materiality with physical limitations, their output a relic of a moment that will never happen again; digital tools do away with these limitations, trading perfect materiality for clean and sometimes made-up infinity.

Making a Choice

The analog vs. digital question can be treacherous. Choosing the analog route to embrace a narrative of quality materials and vintage mechanisms — before even considering what the tool should do — is risking designing a lifestyle product bound to sit on a shelf rather than becoming a great creative tool. Conversely, assuming from the outset that the latest technology is necessarily superior in all aspects risks overlooking important philosophical implications and ultimately produces a technical specification rather than a design.

Designers do not always have a choice — or their options may be constrained. Sometimes, a truly analog tool can be more complex or expensive to build than a digital one — especially software, which by definition is always digital. Finally, a creative tool can be both analog and digital at the same time. In electronic music, for instance, some synthesisers offer digital sequencers with convenient workflows alongside analog synthesis. A pen plotter can trace complex paths that have been generated digitally, using real ink on paper. This presents interesting routes to explore for future tools, where we may see new hybrids; for instance, a portraitist could save actual *paint presets* that a palette could later mix perfectly, taking into account light variations throughout the day, correcting 'white balance', and adjusting for humidity levels in ways that the human eye cannot.

Concepts

This section lists common ingredients designers should consider when moving from high-level ideas to implementation.

Some of these concepts, like the *Node Graph*, form the basis of distinctive tools such as Substance 3D Designer and Houdini, with which they have become synonymous. Others, like *Modes*, are more common, almost unnoticeable, and yet they play an important role in the background. While some relate specifically to software, they all have a wide range of applications and, crucially, can be combined.

This can serve as a basis for critically examining the archetypes we take for granted. The most interesting ideas often come from breaking with established ways of doing things — or mixing them up. *Clip Launching*, for instance, has made music-making more intuitive and spontaneous than it was when the linear *Timeline* was dominant.

Abstraction

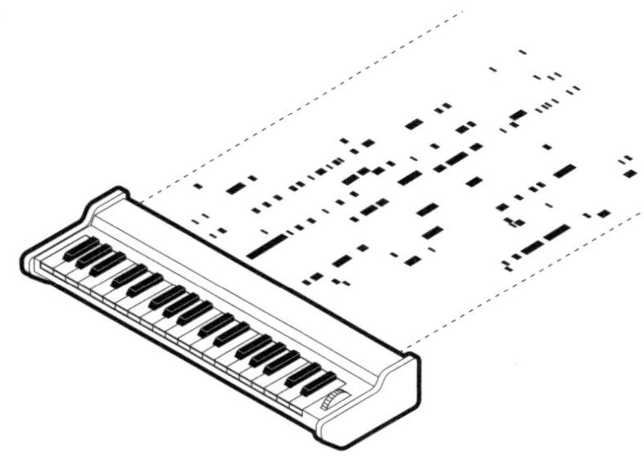

Long before we learned how to capture them with microphones, musical pieces were recorded by hand on annotated sheets of paper, enabling the preservation and transmission of small ideas and works of genius alike. In digital music tools, a similar concept exists, called MIDI (short for Musical Instrument Digital Interface). It allows not the recording of transmitted sound, but rather the timing of notes played, along with their volume and expression, so that the pattern can later be played automatically by another instrument using the same or a completely new sound. Perhaps less obviously, 2D and 3D vector graphics operate in a similar manner: shape descriptions are stored in an abstract form, allowing reproduction in new artworks and, among other things, the indefinite modification of line colour and stroke thickness, without losing the original intent.

Such concepts have been widely adopted in digital workflows, where they are highly valued for enabling reuse and duplication of ideas, non-destructive experimentation and rapid iteration, and the generation of multiple new derivative outputs. In physical form, whilst less flexible, designers may consider using a compass to draw circles of any scale.

Clip Launching

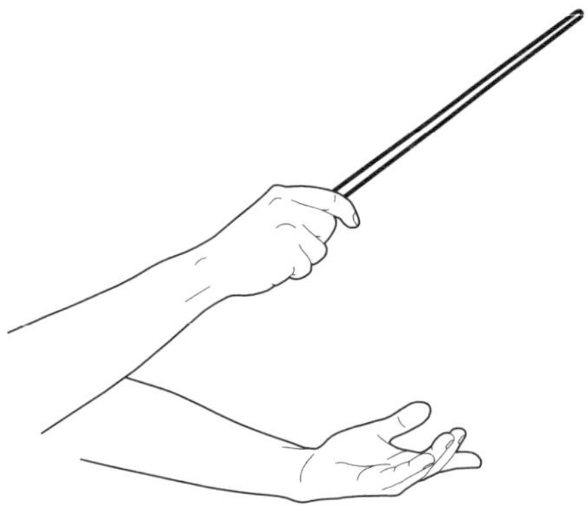

When the musicians of an orchestra play their part under the instructions of the conductor, the piece has already been composed, and they each know exactly what they have to do. However, the conductor decides when things happen and at what tempo, among other things. The concept of clip launching is similar: after positioning the musical or visual elements to be played, usually on a grid layout, the musician–conductor triggers them at different times, individually or collectively. The music production software Ableton is a well-known example of a creative tool that uses clip launching. It is also increasingly present on digital DJ decks.

Clip launching is well-suited to nonlinear work and live performance, where the ability to adapt to the flow and follow inspiration is essential. For composition, some artists find it helps with a more liberated form of ideation, while others find it encourages a certain messiness, whereas, by contrast, a more traditional linear workflow would force them to make early decisions.

Effects and Filters

The concept of effects — *FX* — encompasses everything that allows for an alteration, temporary or permanent, of the object of creation. In filmmaking, we speak of *visual effects* — or *VFX* — ranging from green-screen background replacement to simulated water, fire, and smoke, and digital makeup or face replacement. In music composition and performance, typical effects include adding reverb or echo to make the sound feel more spatial. In photography, especially in smartphone photography, effects are often referred to as *filters* and can range from emulating the colours of film photography to adding — or compositing — new objects into the scene with the subject. Effects are often applied as a final creative touch when everything else has been addressed.

While widespread use across almost all creative tools speaks for itself in terms of both usefulness and popularity, effects can jeopardise the creative experience by imposing a pre-canned formula rather than encouraging experimentation and individual expression. When their role becomes predominant in the tool, the long-term risk is the uniformisation of the creative output.

Instancing

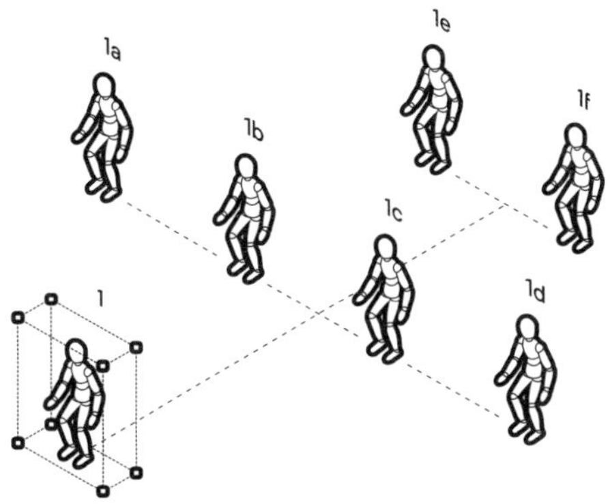

The ease of *copy and paste* is one of the most universally appreciated features of digital tools. Instancing extends this concept by maintaining a dynamic link between the original and its copies. In a 3D modelling tool, an artist can create instances of an object while still working on it; each subsequent change to the original will automatically propagate to every copy as it occurs. This could be used to create a field of millions of modelled flowers, particularly useful when coupled with a parametric workflow to add subtle variations in size, colour, and orientation to each instance. Instancing also exists in 2D illustration, music, and game engines under various names such as *colour swatches*, *styles*, *collections*, *components*, or *prefabs* (prefabricated parts). Sometimes the original is not created by the user but provided as part of a library, ready for selection and reuse.

While a powerful tool, instancing can be difficult to grasp because the dynamic link between the original and its instances is often invisible, leading to unexpected behaviours when editing.

Keyframing

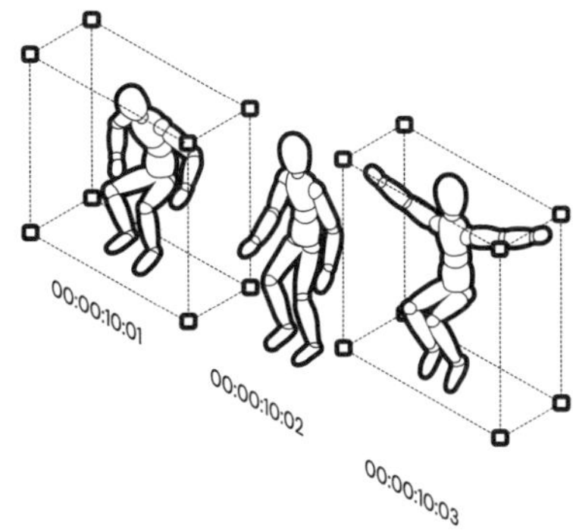

Keyframing is arguably one of the best-kept secrets of animators, removing the need to define each frame of an animation in its entirety, as was once required (for example, a Disney artist painstakingly hand-drawing every picture). In software applications such as After Effects and Blender, keyframing allows the setting of an object's position at key moments in time, namely at the beginning and end of a movement path. The intermediate positions are then computed to enable the object to move automatically during playback. Any parameter can be animated this way, and keyframing is also used in music production, for example, to modulate volume level or an audio effect over time — a process often referred to as *automation*.

While simple to get started, keyframing can be difficult to master and requires an understanding of timing and physics concepts, such as acceleration curves and ease-in and ease-out, to avoid robotic, rigid-looking results. In music, some artists prefer recording hand-operated effect modulations rather than keyframing, as a direct human input can appear less planned and feel more expressive and authentic.

Layering and Mixing

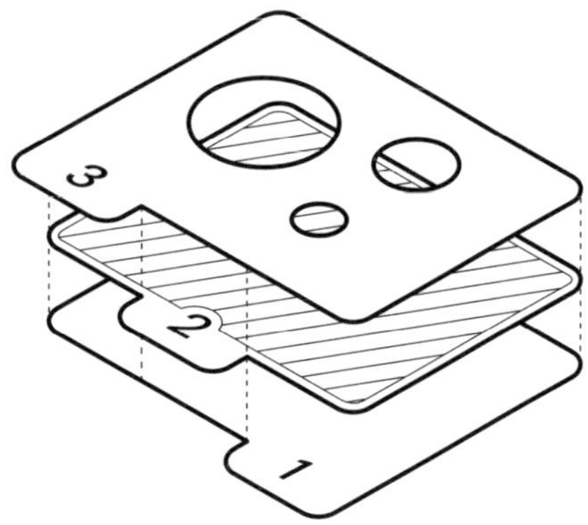

Layering combines multiple individual elements into a single combined output. In the real world, this could be achieved by stacking transparent sheets of paper, each with a different drawing, or by playing two instruments simultaneously. In digital imaging, it was popularised by Photoshop, but it is also possible in video editing and other software applications, where multiple clips or timelines can be stacked. Audio mixing is a form of layering in which multiple tracks are played simultaneously. The primary control for the user is the opacity of the layers, or, with audio, the loudness of each track.

While layering and mixing exist in both the physical and digital worlds, digital media do not exhibit the obvious physical drawbacks, such as reduced visual clarity or increased noise. Beyond combining, layering and mixing are also used to test new ideas quickly, for instance by duplicating an existing layer or track rather than altering the main one. Consequently, the main drawback of layering and mixing is that they can encourage messy structures, leading to mistakes and increasing file size.

Modes

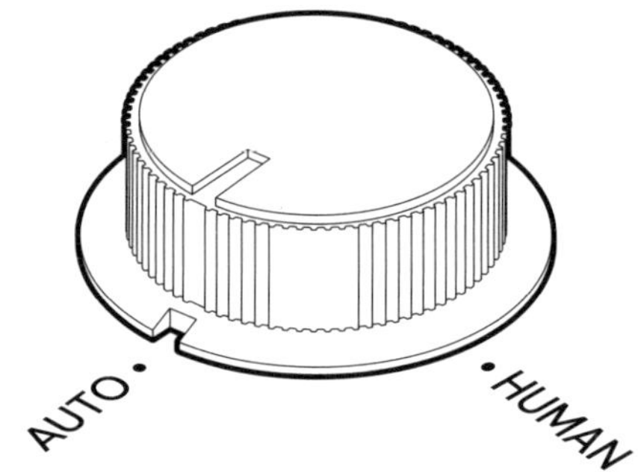

Modes allow users to change a tool's behaviour by switching between states. Each state — or mode — may use a different interface and concept corresponding to a specific stage in the creative process. A digital camera has two main modes: *capture*, in which the camera can be used to take pictures, and *review*, in which the camera displays previously taken pictures. Resolve, a popular video editing software, offers modes for each stage of the film editing process, from cutting footage to adding 'VFX', 'colour grading', 'sound editing', and exporting the final movie.

Modes are useful for reducing perceived complexity by ensuring that the appropriate tools or concepts are available only when needed. When well implemented, they can also help users adopt a more structured workflow. However, modes can be disorienting if users cannot figure out how to switch between them or, worse, are unaware of their existence.

Modularity

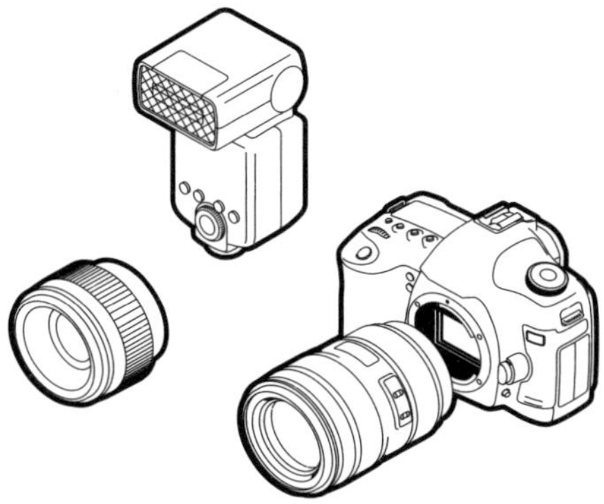

In creative tools, modularity enables adaptation to better meet the needs of specific users and to create more efficient or entirely new workflows. A prepackaged paint kit, with combinations of colours and brushes, is an example of an early modular tool. Cameras are often designed to be modular, allowing photographers to combine different lenses (with varying imaging properties), a camera body, and accessories such as a flash. Modular synthesisers consist of multiple sound units, possibly from different manufacturers, connected via cables (*patches*) to produce unique musical patterns. In software tools, modularity often takes the form of custom menus and window arrangements that users can populate with UI elements corresponding to the functions and workflows they use most. A node graph is also a form of modular system.

While modularity is unavoidable in some tools, it can introduce added complexity. As such, it becomes risky when used as an excuse to avoid design decisions, especially when the audience includes beginners and non-specialist users.

Node Graph

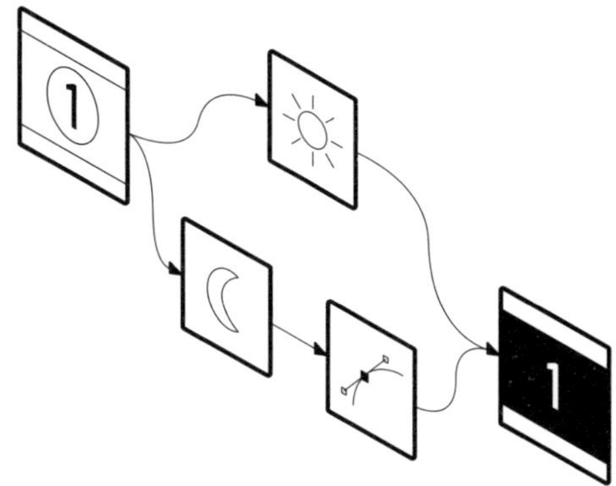

An alternative to programming by writing instructions, a node graph enables the visual mapping of instructions. Each node (represented as a box) executes a specific prebuilt instruction, and its inputs and outputs can be connected to other nodes (represented as lines). In the film-editing software DaVinci Resolve, colour grading can be performed using a node graph. Typically, the first node represents the original movie clip (input), and the last node represents the same clip with its final appearance (output). In between, the colourist places nodes to adjust contrast and saturation and to perform various image manipulations. In the 3D modelling software Rhino 3D, complex forms favoured by architects, such as those working at Zaha Hadid Architects, are generated using the Grasshopper node-graph interface. Many generative AI tools have adopted node graphs as an alternative to chatbots to provide users with greater understanding, flow, and control.

Node graphs are usually preferred for parametric tools, where the effect of each node and connection is visible in real time. While simpler to visualise and learn, they can be difficult to scale for more open-ended programming.

Parametric

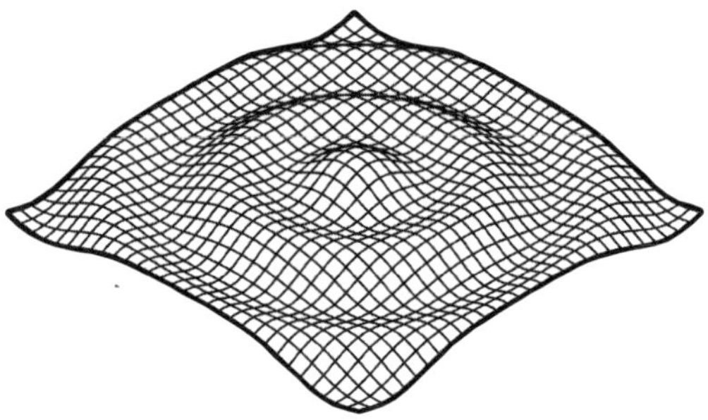

There are times when, instead of building something *brick by brick*, it may make more sense to define rules, driven by just a few controls — or parameters — that can build it *parametrically*. In this way, multiple entities can be constructed, and mathematics can perform tasks more efficiently and accurately than a human hand. Sometimes, defining the rules is part of the creative goal. Most 3D software applications support parametric design, either via a node graph, as in Substance 3D Designer, or via programming, as in Cinema 4D. As for physical tools, modular synthesis is a form of parametric music making in which programming occurs through the control and routing of electrical signals.

Parametric creation is often the only way to produce large-scale works with pseudo-realistic variation driven by mathematical progression and noise. While some appreciate the surprises it brings and the seemingly infinite possibilities, others may find it too formulaic and overtransformative, in that it does not allow for direct human expression or genuine natural variation.

Programming

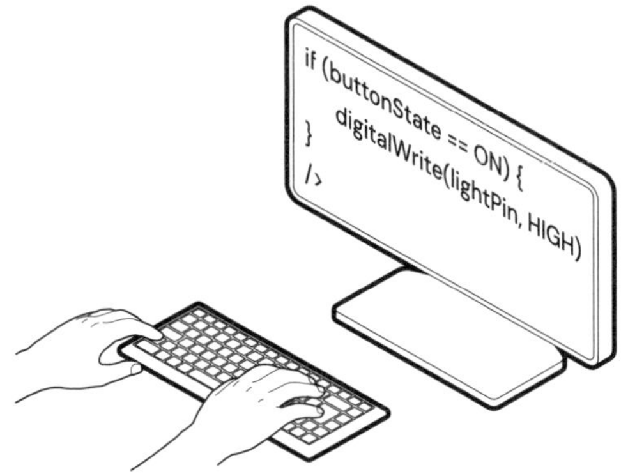

In brief, programming is writing a set of instructions that dictates the behaviour of an object — physical or digital — to which the program is attached. In creative tools, programming is rarely the only concept or interface. More likely, it is a special feature for advanced users who create their own dedicated behaviours within the tool. For instance, in 3D modelling software such as Cinema 4D, one might write a program — or script — to generate a complex animation using mathematics, rather than doing it by hand, frame by frame. Sometimes, as with Arduino microcontroller boards, there is no alternative to programming. Learning to do it often becomes part of the fun and community engagement.

Programming is open-ended and almost limitless, meaning users can create things that go far beyond the use cases the tool's designers imagined. However, it requires strong logic, abstract thinking, and a methodical approach that some people find hard to learn and apply.

Rendering

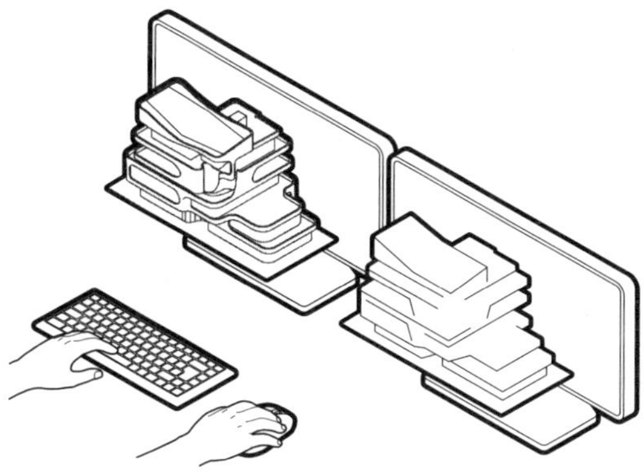

Unlike painting or sculpture, where the result of the work is available to be seen at all times — whether a piece is finished or still in progress — some creative processes require a separate step called *rendering* — or *development* — before the final artefact can be enjoyed. This is the case with photography, digital illustration, 3D computer-generated art, and modern music production. Beyond the technical requirements of exporting media in a specific format, rendering is where some artists feel they have the greatest freedom to express their creativity. In fact, software tools such as Lightroom for photography are largely dedicated to rendering. In 3D, the skills and talents required for modelling objects can differ substantially from those required to create a beautiful final image through rendering.

A distinctive aspect of creative processes with a rendering phase is that multiple outputs can be generated from a single piece of work. In photography, we speak of *developments*; in music, of *mixes* and *remixes*; and, more generally, of *versions*. While it can cause a form of choice paralysis, it also allows artists to revisit past work with a new set of eyes — and refined expertise.

Sampling

Traditionally, the base materials used in creation have been largely natural or derived from natural phenomena: paint from pigment; textiles from animal hair; sound from vibrations; and photographs from the concentration of light on chemically treated film. In contrast, since the rise of collage in the 1910s and, perhaps on a more noticeable scale, sampling in hip-hop music in the 1980s, the base materials used for creation have increasingly been man-made artefacts. As Roger Linn puts it in the interview chapter, sampling allows for artists 'standing on the shoulders of giants' by repurposing small pieces of existing art and music into their own work.

While well established in music, where it has enabled the emergence of popular new genres and led to concepts such as track remixing, sampling is often marginal in other forms of human expression. Beyond the legal challenges it may pose regarding copyright, it is sometimes perceived as cheating, especially if the source artwork is already highly regarded — or if the final new work is judged insufficiently altered in form.

Skeuomorphism

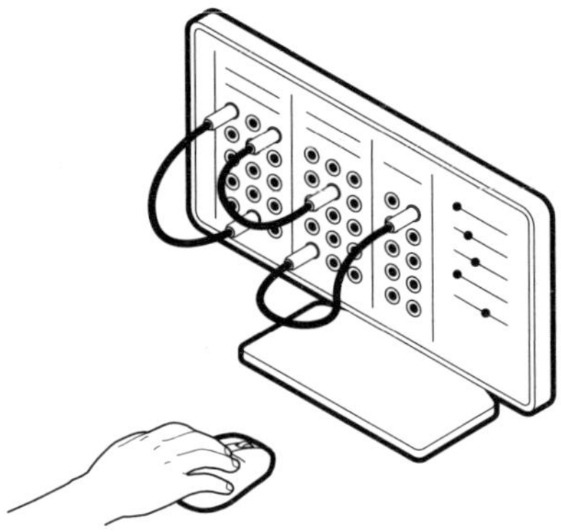

In the history of digital interfaces, skeuomorphism is a recurring theme. In theory, it posits that by making a digital interface look — and behave — in a way similar to a well-known physical interface, users are expected to use it more intuitively than a brand-new concept. In practice, there are few universally known physical interface concepts beyond simple ones that use knobs, buttons, and switches. Reason, by Reason Studios, remains the most ambitious and successful digital application that uses skeuomorphism to guide the entire user experience, in which music-making incorporates the familiarity and possibilities enabled by connecting virtual instruments with virtual cables, as if it were all happening in a real-world studio.

While it often facilitates familiarisation and can introduce a degree of playfulness, skeuomorphism is always on the verge of imposing unjustified rules and complexity on interface challenges — and new possibilities — that are increasingly detached from real-world references.

Timeline

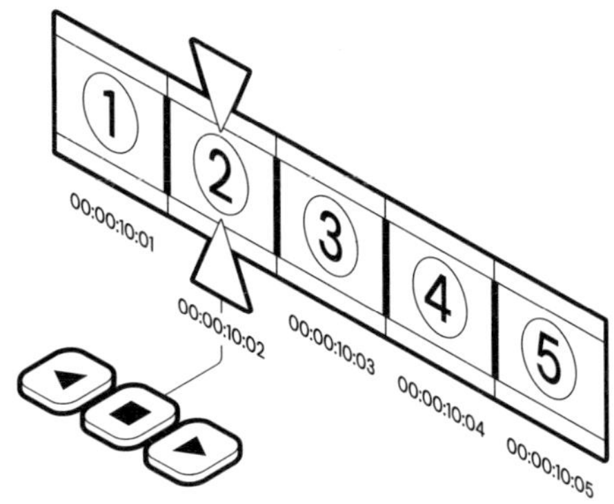

There is a good reason why the timeline has become a fixture in the majority of digital tools used for making music and videos: it is a powerful way to represent information that changes with time, as a *clip*, usually a long rectangle on which a marker — or playhead — moves from left to right when the user presses *play*. Inspired by the physical tape reels that musicians and filmmakers used to manipulate and edit their work, the timeline, and the digital clips on it, maintain their order at all times. Artists can still cut, delete, move, and rearrange clips, and zoom in to do so with precision. They can also name, colour, and duplicate clips to improve visual organisation.

Timelines work well in creative tools for composing and producing linear work, such as traditional videos, films, and music tracks. However, if the ambition is interactive (for instance, a video game or a live musical performance), other concepts such as *Clip Launching* — or even *Programming* — might be more appropriate.

3D Scene View

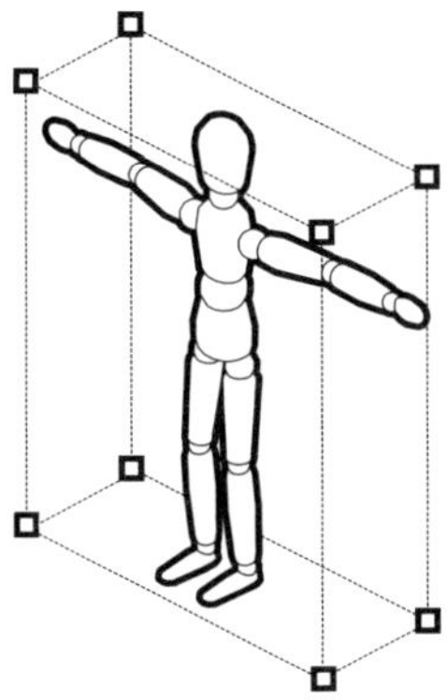

The 3D scene view — or just *scene view* — is the default representation of content created and manipulated in software applications that use three dimensions, such as 3D modelling tools like Blender or game engines like Unreal Engine. It allows the user to have an overview of a virtual world, which may include as few as one virtual object, usually the one being modelled, or millions of them, including characters, components, and props that make up a scene. A good real-world comparison is a movie set. In fact, one of the most common uses of the scene view is to place a virtual camera and lights to define the framing and look of a generated output image, usually called a *render*.

While a scene view is often preferred for 3D navigation and direct manipulation, it is not always well-suited as scale or complexity increases. This is why it is almost always complemented by a semantic view — a structured description of the scene — that enables access to otherwise invisible attributes such as relationships between objects and programmed behaviours.

[1] Randall Munroe, *What If? 2: Additional Serious Scientific Answers to Absurd Hypothetical Questions* (London: John Murray, 2022).

[2] Munroe, *What If?* 2, 47–50.

[3] Richard Godwin, '"More Are Published Than Could Ever Succeed": Are There Too Many Books?', *The Guardian*, March 21, 2025, theguardian.com/books/2025/mar/21/more-are-published-than-could-ever-succeed-are-there-too-many-books.

[4] Mihaly Csíkszentmihályi, *Creativity: Flow and the Psychology of Discovery and Invention* (New York: HarperCollins, 1996), 114.

[5] Csíkszentmihályi, *Creativity*, 114.

[6] Csíkszentmihályi, *Creativity*, 114.

[7] 'Product', *Cambridge Dictionary*, accessed January 8, 2026, dictionary.cambridge.org/dictionary/english/product.

[8] Bill Moggridge, *Designing Interactions* (2006; repr., Cambridge, MA: MIT Press, 2007).

[9] Moggridge, *Designing Interactions*, 244–47.

[10] Moggridge, *Designing Interactions*, 149–50.

[11] Bill Moggridge, *Designing Interactions* (DVD), chap. 4, 'Adopting Technology', 39:30.

[12] Juvenal, *Juvenal and Persius*, trans. G. G. Ramsay, Loeb Classical Library (ca. 100–127 CE; repr., London: William Heinemann, 1918), Satire 10, lines 77–81.

[13] Bill Moggridge, *Designing Interactions*, foreword xii.

[14] Brenda Laurel, 'Interface as Mimesis', in *User Centered System Design: New Perspectives on Human-Computer Interaction*, ed. Donald A. Norman and Stephen W. Draper (Hillsdale, NJ: Lawrence Erlbaum Associates, 1986), 67–85.

[15] World Economic Forum, *The Future of Jobs Report 2025* (Cologny/Geneva: World Economic Forum, 2025), 21.

[16] Donald A. Norman, 'Human-Centered Design Considered Harmful', *Interactions* 12, no. 4 (July 2005): 14, doi.org/10.1145/1070960.1070976.

[17] Norman, 'Human-Centered Design Considered Harmful', 16.

[18] Don Norman, *Design for a Better World: Meaningful, Sustainable, Humanity Centered* (Cambridge, MA: MIT Press, 2023), chap. 22, 'Moving from Humans to Humanity', par. 2.

[19] Daniel Kahneman, *Thinking, Fast and Slow* (New York: Farrar, Straus and Giroux, 2011), chap. 25, 'Bernoulli's Errors'.

[20] R. W. Brunskill, *Illustrated Handbook of Vernacular Architecture*, 4th ed. (1971; repr., London: Faber and Faber, 2000), 27–28.

[21] Gilberto Duarte Carlos, Mariana Correia, Sandra Rocha e Sousa, and P. Frey, 'Vernacular Architecture?', in *Seismic Retrofitting: Learning from Vernacular Architecture*, ed. Mariana R. Correia, Paulo B. Lourenço, and Humberto Varum (London: CRC Press, 2015), 11, doi.org/10.1201/b18856-4.

[22] Carlos et al., 'Vernacular Architecture?', 11–16.

[23] Herbert Spencer, *Principles of Biology*, vol. 1 (London: Williams and Norgate, 1864), 444.

[24] Adam Smith, 'Our photos and emails are silently releasing tonnes of carbon, new report says', *Independent*, October 26, 2021, independent.co.uk/climate-change/sustainable-living/photos-emails-carbon-climate-change-data-b1945701.html.

[25] 'Design the career of your dreams,' Adobe, accessed January 10, 2026, careers.adobe.com/us/en/design-team.

[26] Louis Henry Sullivan, 'The Tall Office Building Artistically Considered', *Lippincott's Magazine* 57 (March 1896): 408.

[27] Adolf Loos, 'Ornement et crime', *Les Cahiers d'aujourd'hui*, no. 5 (June 1913): 247–56.

[28] Garson O'Toole, 'Good Artists Copy; Great Artists Steal', Quote Investigator (blog), 6 March 2013, quoteinvestigator.com/2013/03/06/artists-steal/.

[29] Natalia Ilyin, interview question by authors: What do you think motivates people to use certain tools?; see page 76.

[30] Constantine P. Cavafy, 'Ithaka', in *C.P. Cavafy: Collected Poems*, trans. Edmund Keeley and Philip Sherrard, rev. ed. (Princeton, NJ: Princeton University Press, 1992), 36.

[31] Anna L. Cox et al., 'Design Frictions for Mindful Interactions: The Case for Microboundaries', in *Proceedings of the 2016 CHI Conference Extended Abstracts on Human Factors in Computing Systems* (May 2016), 4, doi:10.1145/2851581.2892410.

[32] Viktor Lowenfeld and W. Lambert Brittain, Creative and Mental Growth, 8th ed. (New York: Macmillan, 1987), 191–92.

[33] U. Frey and R. Morris, 'Synaptic tagging and long-term potentiation', *Nature* 385, no. 6616 (1997): 533–36, doi.org/10.1038/385533a0.

[34] Scott Belsky, 'An Interview with Adobe Chief Strategy Officer Scott Belsky', interview by Ben Thompson, *Stratechery*, podcast, April 13, 2023, 'Adobe and the Web', stratechery.com/2023/an-interview-with-adobe-chief-strategy-officer-scott-belsky/.

[35] Stephen Farrugia, 'The Interface and the Potential', *Faster and Worse*, December 21, 2022, fasterandworse.com/the-interface-and-the-potential/.

[36] Farrugia, 'The Interface and the Potential.'

[37] Stephen King, *The Stand*, rev. ed. (London: Hodder and Stoughton, 1990), chap. 64.

[38] Howard E. Gardner, *Multiple Intelligences: New Horizons in Theory and Practice* (New York: Basic Books, 2006), 'Chapter 8: Multiple Entry Points Toward Disciplinary Understanding,', 'A Word About Special Populations'.

[39] Howard Gardner, *Frames of Mind: The Theory of Multiple Intelligences* (New York: Basic Books, 1983).

[40] Gardner, *Multiple Intelligences*, 'Chapter 1: In a Nutshell'.

[41] Gardner, *Multiple Intelligences*, 'Interactions Among Multiple Intelligences', 'Bottlenecks'.

[42] Nicholas Carr, 'All Can Be Lost: The Risk of Putting Our Knowledge in the Hands of Machines', *The Atlantic*, November 2013, theatlantic.com/magazine/archive/2013/11/the-great-forgetting/309516/.

2 | Interviews with Practitioners

What are humans trying to achieve when they design new tools? What is their process, and how did they arrive at it? Frequently, those who design creative tools are not first and foremost formally trained as designers. Instead, they often started designing tools because of an interest in using them to create new things.

This chapter brings together various perspectives — insights from individuals who consider themselves *designers* of creative tools, as well as those who hadn't titled themselves as such. They are musicians, entrepreneurs, painters, tinkerers, writers, and designers; all have spent considerable time in their careers developing creative tools. These include physical and digital tools, tools for the hand, and tools for the mind; tools that have sold in the millions and tools that enabled the creation of artworks on view in national museums.

Throughout the interviews, common themes emerge: the tool as an *active* participant that defines the creative output — rather than a passive object; the paradoxical value of friction and limitations for focus; the — sometimes — vital role of enjoyment; the enduring value of physical interactions and tactile imperfections that capture a *human feel*; and the goal of designing for creative thinking rather than mere productivity. Ultimately, the conversations often converge on the tension between human agency and AI, with some warning that *invisible* tools threaten to hollow out authentic creative identity.

Roger Linn

In his prolific career as a musical instrument designer, Roger Linn has pioneered expressive machines with foundational instruments such as the Linn LM1 and the Akai MPC. They played a major role in the transition to modern beat-making and the use of sampling, ultimately fuelling the music revolution of hip-hop and electronic genres. His influence can be felt in most music production and DJ equipment designs, especially through the use of grid layouts, quantisation and the legendary 'swing' features that make music-making with machines more human.

What role does being a musician, and hence a user of your products, play in your design process?

Having worked as a musician in my younger years of the 1970s, I have experienced the difficulty of using the poorly designed tools of the day. When I'd have a musical inspiration, I'd go into my music room and two hours later, find myself sitting on the floor reading product manuals, trying to figure out how to use the confusing products. I found that frustrating and imagined that I could make a better product, one designed more for musicians than engineers, and that works the way a musician thinks — and how he expects it to work. And that's what I've tried to do.

Why is it that designing creative tools seems to require a more hands-on approach than, maybe, other forms of design?

Compared to business or productivity tools, I think creative artists tend to have a wider variety of personal and unique visions for what they want to create. So a creative tool must include a wide variety of methods of creation that the artist might desire, and designing these methods requires an intimate working knowledge of the artist's workflow and needs.

The problem is that designing a tool with a wide variety of creative methods tends to result in high complexity. So I tend to think a lot about how to organise these methods in a logical and intuitive way in order to end up with an intuitive and fast workflow, for example, using commonly accepted conventions and terms, and having many conversations with the people who will use the product. I believe that with sufficient design skill, it is possible to have both flexibility and simple operation.

What about aesthetics?

I'm a strong believer in the idea of *form follows function*, that designing a beautiful or stylish tool should never compromise its function.

That said, I try my best to work within that constraint to create an instrument that is as beautiful as possible. And that can be hard because my current company, Roger Linn Design, is tiny and I don't have the resources that big companies have to create things like custom-moulded chassis that require high up-front costs. Instead, I try to create beauty on the cheap.

For example, my current LinnStrument musical keyboard is essentially a rectangular sheet-metal chassis. But to give it a more rounded and organic form, I attach 1-inch diameter half-round wood pieces to the two long edges. To get an idea of what this would look like, I found six-foot lengths of these half-round wood pieces at a local lumber yard, a commonly available low-cost wood product used for architectural wainscoting.

Also, I wanted to minimise the margins around the large playing surface so that the product appears to consist almost entirely of the playing surface, but the space required for conventional internal connectors would have increased those margins. So I designed a unique and unconventional method of internal connection, eliminating those bulky connectors and enabling the narrow margins that I desired.

Lastly, I preferred the elegant look of a thin musical instrument and had a goal of one inch total thickness. But one part — a MIDI jack — was a little too tall to achieve that goal, and I didn't want to pay to make a custom shorter jack. So I simply use stock MIDI jacks and cut off the tops to make them fit, yet they are still sufficiently strong that I've never had a problem.

You have said in the past that testing your first drum-machine design with Stevie Wonder made you realise what was wrong with it. How do you draw inspiration from testing with real people?

First of all, just being in the presence of someone like Stevie Wonder is an incredible opportunity for a musical instrument designer. Because of his experience, knowledge, and creativity, he has insights that I will never have. Years ago, I showed him an early prototype of my first drum machine, which presented a visual grid of drums and time steps on a computer screen, a method commonly called *step entry*. But in addition to seeing how difficult it was for a blind musician to use a visual interface, I also noticed that it forced the musician — someone who can easily tap out a rhythm in real time — to do a mental translation of his intended rhythm into a mathematical sequence of time steps. This made no sense.

So I went back to the drawing board and came up with a method of tapping in a rhythm in time to a metronome, thereby eliminating this mental translation. But this created a new problem: what if a musician's timing when tapping buttons is not so good? That's when I invented timing correction, which people now call *quantise*, automatically nudging the musician's taps onto the nearest perfect 16th or 8th note. Then I realised that quantised timing can sound stiff, so I came up with the idea of *swing timing* — which I called *shuffle* at the time — by delaying every alternate 16th or 8th note by a variable amount, thereby adding humanness.

Musicians appreciated that the machine adapted to the human instead of the other way around.

The bottom line is that I try my best to listen to what musicians tell me, and also to watch how they work in order to learn what they may not be saying.

Is your goal to humanise the creative process?

Yes, that's fair to say, and in two ways. First, to make the machines easier for humans to use. And second, to make it easy to produce human-sounding music. To my ears, many of the early electronic music sequencing products of the 1970s produced stiff and robotic-sounding music. I wanted to create machines that could help make music that got closer to the human feel of the records I loved, made by talented musicians whose skill and nuanced subtlety made these recordings ideal examples of *the triumph of the human spirit*.

However, one thing surprised me when I made my first drum machine in 1979: the music that many people were making with it often sounded robotic and soulless. This is perhaps because my new concepts, like swing timing — and multiple buttons for playing a drum at different loudness levels — were unfamiliar to many. So they tended to use the product in the simplest way without using these new features. This led me to the realisation that musicians often use music products in the simplest way without digging into the more creative features. In a sense, they tend to make the music that the machines want them to make. If it is easiest for the machine to make robotic music, people will make robotic music, with stiff rhythmic parts, rigid note sequences and little dynamic variation. It was this observation that led me to develop touch-sensitive drum pads on my subsequent Linn 9000 product, as well as a feature called Note Repeat: holding the pads while varying pressure would produce repeated 16th or 8th notes with a natural dynamic variation, one that would be difficult to achieve by playing individual notes on the pads.

But I also noticed something else. It was becoming a popular style to make music that sounded robotic, something I never would have guessed, especially after having created features *intended to help create human-sounding parts*. This taught me that it can be difficult to predict how people will use the tools that I make. I like to say that as a designer, I can

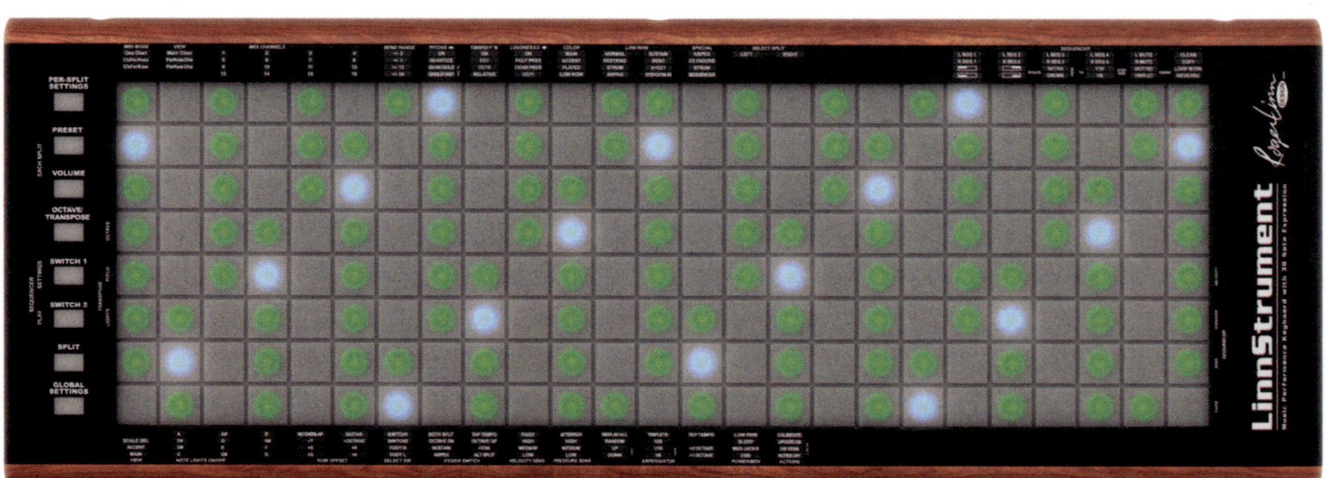

▼ *LinnStrument MIDI Controller (2014)*

create the brush, but the artist decides what to paint with it. I take my best shot at that, but I've been wrong much of the time. So if I can make a tool that's flexible enough for them to take in different directions than I intended, I feel I've succeeded.

What have you wanted to achieve that isn't yet possible?

My LinnStrument product is highly sensitive to light touches, but I'd like to see even higher sensitivity. For example, it would be nice for an electronic instrument to be so sensitive that it produces a different sound when touched lightly with the flesh of your finger or your fingernail, just as with a small acoustic drum. This is tricky because it's difficult to make an electronic instrument sensitive to such light touches without also being sensitive to harsh environmental conditions like heat or humidity. But it's important because I feel that so much of the beauty in music is in that place between silence and whispers.

Also, I think AI will be helpful in making music products easier to learn and use. Today's music products often require studying manuals, which is a bore and assumes that the musician must learn to think like the product. One way of thinking about AI is that it permits changing from the paradigm of the human adapting to the machine to the paradigm of the machine adapting to the human. Already, ChatGPT and Gemini are pretty good at answering questions about music products, and they're getting better fast.

How do you choose between innovation and familiarity?

I don't think they're mutually exclusive. With sufficient design skill and understanding of what the musician wants, a designer can create a product that presents innovative concepts in a familiar and simple way. Albert Einstein once said, 'If you can't explain it simply, you don't understand it well enough.'[1] Similarly, if the designer doesn't clearly understand why the feature is important to the musician, he won't be able to design it in a way that he will be able to easily use it.

When I look around at current music synthesiser designs, I don't see a lot of *design* but rather a lot of *more and better* versions of what existed before — more oscillators, more filters, more effects. This is helpful for musicians who think like engineers, but not so good for musicians who think like musicians. Interestingly, they are giving techie musicians something *familiar* — more of the same engineering-focused tools that musicians have been dragged kicking and screaming to learn for years — but not very innovative for non-techie musicians.

Ultimately, I think what the musician wants from a sound synthesiser is the ability to describe the sound that he hears in his head, then for the synthesiser to create that sound to be played from his instrument. I don't think he wants to adjust filter frequency, resonance, or modulation index. I think AI can help with this. In the same way that AI can create a finished piece of music from a text description, I'd like to see a MIDI synthesiser that can create a new synthesiser sound from a text description in musical terms, or perhaps also from mouth noises, audio recordings or references to existing music, resulting in the desired sound that can be played from his instrument. There are some early attempts at this, but the technology isn't there yet. Perhaps training the AI on lots of synthesis forums will help it better understand how musicians think and speak, as well as the associations between their words and the audio recordings in their forum posts. If the tool is sufficiently innovative in terms of ease of use, perhaps it doesn't have to be familiar.

Your designs were pivotal in the development of hip-hop music, in part because they enabled the easy incorporation of samples of full musical phrases and their recombination into new songs. Today, such techniques have spread to almost every musical genre. Why do you think it remains so popular?

It's popular because it allows people to assemble a compelling musical work without having to learn how to play a particular instrument. I call this *object-oriented composition* or *OOC*, the process of creating a new musical work by combining pre-made loops, beats or sequences in a new and creative way, not unlike artistic collage. Hip-hop does this, as do clip-launching DJs. Given a century of existing recordings to draw the objects from, they shift the focus from creating each and every musical note to a higher conceptual focus of overall composition and arrangement of these objects. Though the resulting music can be limited in detail or musical variation, it's faster than years of violin lessons. In a sense, they are standing on the shoulders of the giants who came before.

How do you anticipate the creative journey of the people who use your instruments in your design process?

I try to deeply understand the musician's needs, even if what they really need isn't what they say they need. Generally, I find that musicians ask for more and better versions of what already exists. There's a nice quote that is generally attributed to Henry Ford, though there is no record of him actually saying it: 'If I had asked people what they wanted, they would have said faster horses.'[2]

Instead of *more and better*, I try to figure out what the musician's true end goal is. Once I have that, I try to come up with a way to achieve that goal that causes him to say, 'I wouldn't have thought of that, but it's exactly what I need.'

Here's an actual quote from Henry Ford that I like even better: 'If there is any one secret of success, it lies in the ability to get the other person's point of view and see things from that person's angle as well as from your own.'[3]

Another point is that with the best of intentions, I'm often surprised by how people use my instruments in ways that I wouldn't have anticipated. So I try to make them flexible enough for that. For example, in my LinnStrument keyboard, I released its software as open source, allowing anyone to change it from what I intended into what they prefer it to be. I also released engineering documents for much of the hardware of the product, allowing people to customise the physical interface to their liking.

How does the development of computing influence your thinking about music-making and the design of musical instruments?

I got my first computer around 1976 at age 21, shortly after Intel released the revolutionary 8080 microprocessor chip. Bill Gates and Steve Jobs used this chip to try to put a computer on every desk, but I wanted to make music

▲ *Linn LM-1 Drum Computer Brochure (1980)*

products with a computer chip inside, dramatically increasing the products' capabilities.

This was an exciting time because young visionaries were putting microcomputers inside everything — cars, phones, calculators, video games, medical devices, kitchen appliances and more. There is a book by Malcolm Gladwell called *Outliers*,[4] which has a chapter on people born in or around 1955, like Steve Jobs, Bill Gates, Eric Schmidt, and a long list of tech titans. I was born in 1955. The point that Gladwell makes is that if you were born around 1955, then when the microchip revolution took off in 1975, you were 20 years old. You weren't still stuck in high school, but you weren't already married with kids and having to play it safe. It was the perfect age and time to leave university and start a company, as Jobs and Gates did. So while I think I was talented, I also had the luck of timing.

My first product was the 1979 LM-1 Drum Computer, the first drum machine with both sampled sounds and a useful real-time recording workflow that corrected your timing errors and added swing timing, all enabled by having a computer chip inside. People were astounded. It reminds me of a quote by Arthur C. Clarke: 'Any sufficiently advanced technology is indistinguishable from magic.'[5] When people would push one of its buttons and instantly hear a real recorded drum sound play, it felt magical to them.

Today, computers are inside most products, and many of those products are confusing to use and learn, unreliable, buggy and quickly obsolete. The new revolution is AI, and I find that people are increasingly ignoring product manuals and instead asking ChatGPT or Gemini for answers. I suspect it won't be long before companies are building AI agents into their music products, allowing people to simply ask their products to do what they want to do in musical terms. I'd like to see that happen, because there's a huge number of non-techie musicians whose music could benefit from music products, if only the products would allow them to think like a musician instead of like an engineer.

Where do you see the future of musical instrument design?

I think we will see more instruments that answer the question, 'What is a musical instrument in the 21st century?' as opposed to guitar, keyboard, wind or bowed instruments that were designed around the mechanical constraints of a few centuries ago.

I find it somewhat odd to see people still playing electronic piano keyboards. Why? Because the piano keyboard — a single, long row of keys — was designed a few centuries ago around the need to place a string behind each key, yet electronic keyboards have no strings.

The guitar is a wonderful instrument, but it has similar problems. You can't easily play accompaniment and melody simultaneously, it takes both hands to play anything, the notes are spaced far apart at one end and too close together at the other, you are limited to playing only one note at a time on each string, and the dots on the neck stay in the same place regardless of which musical key you're playing in. Most people would say 'That's just the way it is,' but modern instruments like my LinnStrument solve these limitations, yet enable the same performance gestures like vibrato, pitch bends, and continuous control of each note's loudness and timbre.

The essential problem with acoustic instruments is that the human interface is limited by the need to make acoustic sound. Electronic instruments solve this problem by generating the sound electronically, so the human interface can be optimised for the human, and the sound generation can be independently optimised for generating sound.

Why do people still play these arguably obsolete musical interfaces? Tradition? A sense of nostalgia around music and musical instruments? Probably both. But also, humans take a long time to learn a musical instrument, and I suspect that many people are afraid of taking that time to learn a new instrument if the company making it may not exist next year, yet that company may not exist next year if no one buys their instrument.

A last thought, one that is particularly exciting for me, is the current availability of simple-to-use, low-cost electronic, software and sensor development tools that allow techie musicians to build their own musical instruments. These tools are the modern equivalents of working with wood, frets, tubes, reeds, drums, etc. I co-teach a workshop every year at Stanford on designing instruments, and I love seeing the ideas that students come up with. Who knows what form of musical instrument will be popular in 50 years? My LinnStrument is my best guess, but it will probably come from some young innovator using these creation tools, someone whose ideas just happen to strike a nerve that resonates through the culture of music-making.

Mats Herding Solberg

reMarkable

Since the early days of reMarkable, Mats Herding Solberg has led design for the company behind the world's first 'paper tablet.' Following a breakout 2017 Kickstarter campaign, the reMarkable 1 popularised e-ink as a medium for digital note-taking and sketching. By prioritising a tactile, paper-like feel, the device champions a softer idea of technology and delivers a distraction-free experience designed to foster focused thinking and creativity.

What do you do at reMarkable?

Magnus Wanberg started the company, and I was a founding employee, together with our then CTO* Martin Sandsmark. Magnus called our trio 'the hustler, the hacker and the designer.' I built the design department over many years; now we are roughly 500 employees, and 80 of them are designers, and I am CPO†, so ultimately responsible for our products. I've just changed my title from CDO‡, which I had for ten years.

I think I've been a driving force to push the importance of design, but also the importance of not being a design-driven company, which is a scary fallacy where people go into building pure design teams that become echo chambers. My role all along has been championing unparalleled user experiences that go from the physical side to the software side, to the emotional side. The sum of that constitutes really good product experiences. That is at the core of what designers should do. They should be the tension, always striving for brilliant user experiences, and that can be articulated in multiple different ways. That's not to say that you have to go super premium or that you have to go super wide on your feature offering. It's rather being those guarding the concept of doing a few things really well. What I've been doing a lot over the years is trying to argue that there are tangibles, which can often be a basic business case or a commercial value proposition. And then there are these intangible things that are not necessarily directly measurable in the business case, but that end up enabling valuable user experiences. Today, as the CPO I have the overarching responsibility for long-term product strategy: trying to point us in a direction, create aspiration so that we also

* Chief Technology Officer

† Chief Product Officer

‡ Chief Design Officer

instil enough ambition in the shorter term — because we're reaching for something far — and trying to understand and guide the dynamics of our product portfolio.

Did you have a different perspective on the product, being yourself a designer — and hence a specific kind of user in the first place?

Definitely. But being a designer and having spent a lot of time sketching in my days, it was obvious, pretty early, that we were not making a sketching tool. Wacom Cintiq products already existed, and they were really good. But the level of visual thinking that I do — which is not necessarily the same as sketching — that was a really important element for me, and I think that impacted Magnus and helped shape the product. For instance, in the very beginning, we didn't know which size we wanted to make the reMarkable 1, and we started off with the Moleskine as a benchmark. To me, that was a bit limiting on the visual thinking part. So we actually went for a bigger device when launching the reMarkable 1. Lately, we've launched the smaller reMarkable Pro Move, and I've reflected that I now have different purposes for different tools: the value of the small tablet is its portability, which makes me use it a lot more. That's, of course, a privilege for me as I have an abundance of available paper tablets, but it's still interesting that it started with 10.3 inches and now we've gone both bigger and smaller — even if I still think 10.3 inches is the best compromise!

Do you think you need to be bold to propose a monofunctional tool in a world obsessed with platforms and multifunctionality?

Back in 2017, if you were to pay $600, $700 for

▼ *reMarkable 1 Paper Tablet (2017)*

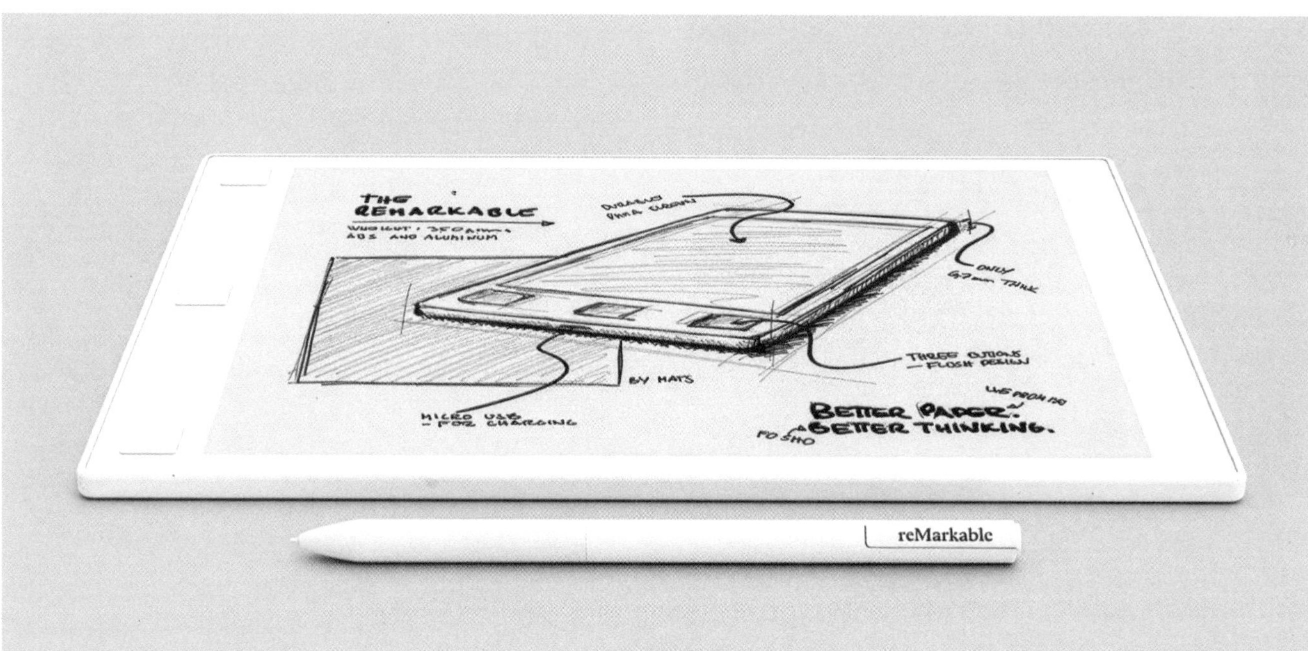

a technology product, it had to do a lot. So making a single-purpose product was pretty bold — yes — and we actually marketed ourselves on it doing very few things. As a designer, this often defines how capable you are because it's very easy, when you're not completely certain about what you're making, to start wanting to add functionality or optionality. I have this instinct myself, but if you're bold enough, you're going properly vertical because you know that the problem is the right problem and then you start saying no to your instinct of broadening — which would only give yourself more things to fall back on by making a horizontal product. Then you'd be kind of mediocre at a lot of things instead of really good at one thing.

Have you witnessed people doing things in a way that made you think: 'This would not have existed as it is without our tool'?

We see people using reMarkables in ways we did not imagine, which is super cool. Limitations and restrictions are often the best friend of a creator. In the early days, the black and white reMarkable tablets were really limited, but we had people on Reddit and Instagram sharing super intricate and complex drawings, which was mindboggling. They loved the challenge of doing it on the reMarkable. For instance, we had one shader tool which was really shoddy to work with and export, so people went through enormous detours to get stuff into Illustrator and continue working on it. But that's also kind of the same as paper. People still default to paper to such a large extent, and there's a reason for that. The fear, but also the value of a blank piece of paper as opposed to whichever tool you start out with, is clear: absolute freedom.

How do you handle feature requests and feedback?

Usually, you have a vocal power user minority, and then you have a silent average — but still happy — user majority, and they're significant. And the typical fallacy is that you should engage more with that 3% minority, where people also hate features that the 97% love. So in the worst-case scenario, you start optimising out of the majority and into the minority. And that's definitely a position you can take. Of course, there is magic in the power user community, and sometimes I really feel that I could do more of that — but then that's not the product that we are building. We try to strike that balance through a combination of listening to vocal users and balancing that with observations. So, a typical example is that you have the Reddit community, and they make it seem like everybody wants this particular feature. You build it, you launch it, and then you actually see if it's being used — or not. But still, you spend effort points that did not go towards the majority and from a socialistic point of view, it's unfair!

How do you define the role of limitations in serving the user's primary purpose?

If you don't dare to remove, you will only bloat, and we're thinking a lot about that. Of course, our products today are still quite simple. We have a very hard principle of trying to avoid choice wherever we can because choice creates friction, and we should remove any type of friction that's not helping you to think — which is basically what we believe people are doing on our tablets. You can have a sensation that you need choice — like a lot of the power users, by the way. They're like 'I would like to be able to customise my brush, particularly for my preference.' And my challenge to them is 'How would that make your thinking better?'

Friction that takes you away from your intent or purpose is not healthy. But friction — and frustration — can be healthy when it forces you into a new mode of thinking or reflecting or creating. When a tiny baby is on the ground trying to figure out how to crawl, it's like 'don't pick her up just because she sounds frustrated.' Because that frustration drives her to achieve something. There's a brilliant quote from a YouTuber that did a One Year with my reMarkable type review. He said, 'What I love the most about reMarkable is the fact that it slows my brain down to the speed of the pen.'[6]

Why is it that designing tools seems to require a more hands-on approach than, maybe, other forms of design?

I think it comes down to the fact that with a tool, there is an intent to solve a problem. Not necessarily an explicit problem, but your intent is to do something in a better way than before the tool existed. You also need to appreciate the problem — you need to feel it in your body. That's obviously a driver for needing to be hands-on.

If you build a hammer, it has to be a brilliant hammer before it moves towards art on the spectrum. It depends on where you enter the spectrum and what problem you are trying to solve. If it's as good on the functionality side, that increases the sense of ownership — and pride — and the quality of the user experience. And that's a tough balancing act. Yes, you may build a stronger identity for yourself as a designer — or even as a brand — but the second that starts trading off on functionality, it's a bit hypocritical.

There is an internet legend that says that you designed an iPhone for a school project — long before it existed. How did it influence your thinking about design, in a world where hardware and software increasingly need to work hand in hand?

Right [laughs]. I grew up in a home with a father who's been a creative director his entire life. So, he's always been an avid Apple user, and I was privileged as a kid to always have the newest Apple computer — or Macintosh at that point in time — at home. I was looking up to my father, really wanting to create the same way he was creating, and I was really inspired by those products because they were shaped so beautifully, and there was so much attention going into that. Then my father started using the PalmPilot and Handspring products. That was kind of his reMarkable of the old days — with the pen and Wacom technology — but I hated those products. I thought they were so uninspiring as opposed to the Apple ones, and I was like, 'there has to be something in between here.' So for me, there was this continuum where something was clearly lacking. It was 2001, I was in 8th grade, and there was an arts and crafts project. The intent was not to create new technology: the actual deliverable was a thing made in wood, but I ended up designing a product which had an Apple interface I had Photoshopped in a tiny scale onto a foam-type body inspired by the PalmPilot. So the size was basically like the first iPhone, and it sat in a dock, and then I wrote a big report on how the interface worked!

Today, we are plateauing in a world of multipurpose products and on the journey up there, a lot of people have built businesses while disregarding the platform, just operating within the frames of what an app* is intended to be. But all the apps that we needed have been made by now.

I think that those who have been able to build products that are vertical enough will survive, and if you want to do that successfully, you need to build software and hardware in parallel. But the difficulty of it is absolute. In theory — and historically — the approaches are very different. Hardware takes a lot of time while software is iterative, so the two road maps are difficult to integrate. But you can create much better products if you're able to do that. The opportunity for us, with a joint road map of intent, is that we can reduce the hardware to the greatest extent possible. For instance, I would prefer, from a sustainability perspective, from a rationality perspective, and from a price perspective, not to put anything into our hardware that the end user does not need. Most people today carry around iPhones with 10x the power that they need and at least 10x the power that they use, but they believe they might want to use it for something at some point in time. We are kind of appreciative of the fact that the reMarkable tablet is not the primary device of any one user out there. I would prefer to leverage as much of that hardware that others are building to make our products better, as opposed to just putting out the same hardware, which would lead users to have double the impact on the world and carry twice as much technology around.

Is the design of the reMarkable tablets Scandinavian?

We talked a lot about that in the early days, Magnus, I and some of the others that came along. Are we a Scandinavian company or are we a Norwegian company? Are we a European company, or are we just a global company? The typical Scandinavian model is to test in your own market, then expand to Europe, then expand to the USA. We flipped that pyramid from the get-go, but we had no idea about the American population and their preferences in tech, for instance. We knew we didn't want to identify ourselves as a Norwegian company, and we knew we didn't want to identify ourselves with Scandinavian design, which is very much like a terminology that's being used for interior and furniture design. But is the reMarkable inspired by Scandinavian design? Yes — the element of functionality over form is an element of Scandinavian design, and that's pretty clear that you don't add detailing where it's non-functional. That is a principle that we have been adhering to in everything we've been doing, both physically and digitally.

Is your goal to merge analog and digital workflows?

To some extent, that is our goal, but I can try to nuance it a little bit because analog for the sake of analog is just opposition to change. To me, it's about respecting

* Short for 'application', a software program designed to perform a specific function for the user, most commonly associated with mobile devices like smartphones and tablets.

► *reMarkable Marker (2024)*

what's valuable about the analog world in our space, and that's the interaction of pen and paper. We talked about the intimidating nature of a blank piece of paper, whilst at the same time, the value it brings. I really want us to be able to retain that. It's about being respectful to what's valuable in the analog world and, then, trying to connect that to the digital world in a thoughtful manner. Our products are definitely technological products. There is nothing analog about them other than the fact that we're moving physical particles as opposed to a light type of thing. But they're built to replicate some of the value of the analog products that are paper and — more importantly than paper, to be honest — the pen. The pen is the extension of your brain for free-form thinking. Then you have a whole lot of digital workflows which other products handle a lot better than we do. But with a lot of the 'platformised' products, when you've done something, and you click *share*, then you have to select where you want to share it, how you want to share it, who you want to share with and then you just add friction upon friction upon friction. So what I set out as a vision for us is *one-click-in, one-click-out*. So, we're not leaning into the digital space and trying to replicate that on the wrong surface, but just making sure that you have a sense of security and confidence that it is sent to the right workflow.

Are you just trying to make people happier?

We've done a lot of research over the years on what motivates people. I would say 'yes', but the question is how we can contribute to that, and that's the value of time spent on certain types of activity that you do on certain types of product. I feel confident that pushing our users to use their reMarkable tablet more — not 12 hours a day but an hour or two — is only good for them. I hope and believe that we can forever say that this is true. That cannot be said of your phone, your laptop or your TV, for that matter. What we see we can deliver in people's lives is the ability to think deeply, and this is kind of the scarcity of today. Multitasking, productivity, everybody's building that, but I don't think it is a good measure of what we do. A lot of the tools that we're using today, a laptop for instance, are good at *busy work*, finalisation and productivity — say the last four iterations of whatever you're creating — but not so great at the first one. And that's our role. It's a place where I really believe we can contribute to making people happier.

What feedback from reMarkable users makes you happiest?

Constructive feedback [laughs]. What really makes me happy ultimately is when I feel that we've made an impact in people's lives. If people feel that they have more time to focus, they're getting more time to think deeper or process differently, then that's the most value I think we can bring to the world. A lot of people are genuinely saying — and I hope they mean it — 'reMarkable has changed my life.' We hear that quite often, and we're not paying people to say it [laughs].

To Domus,* you said you 'set out to design a device' that would empower people to 'use the world's most powerful computer: their mind.'[7] How critical is that mission statement in a world where we hear

---

* Established in 1928 by architect Gio Ponti, *Domus* is an influential Italian architecture and design magazine. Historically, it has served as a critical platform for exploring the intersection of the built environment, industrial design, and the humanities.

that the future is not for people to connect to their mind but rather to delegate their thinking?

You can ask a whole bunch of reMarkable users today, and they will say they want to handwrite and have that as an interface with an AI chatbot type of thing. That's the opposite of what I want us to achieve. Don't get me wrong, I think there's a lot of value in AI. I use it a lot myself, actively trying to learn and understand how it creates value. But what we refer to as human thinking is a completely different mental model from the mental model of a neural network*. Those two are complementary, which means that original thinking will always have a role in guiding AI and making sure that the output from AI is valuable. For productivity, the regurgitating of existing information — or just the repositioning of it — that's something that we are not taking a role in. But I think we have a really important role in protecting the ability of people to think their own thoughts without too much interference or without just defaulting to asking. For school children who are not able to handwrite anymore, it has such a huge impact on their ability to learn, and they've become so used to just seeking information when they need it. They glance at it, but they don't properly process it. In a world where we're not continuously nurturing our ability to think, we will also make the computer more stupid if we don't help it, if we don't guide it to some extent. The speed of evolution on AI currently is insane, and a lot of what I consider 'busy work' — or productivity work — will stop existing to some extent. What is going to remain, I don't know, but I believe human thinking and the process of following your own mind and your own string of thoughts is incredibly valuable.

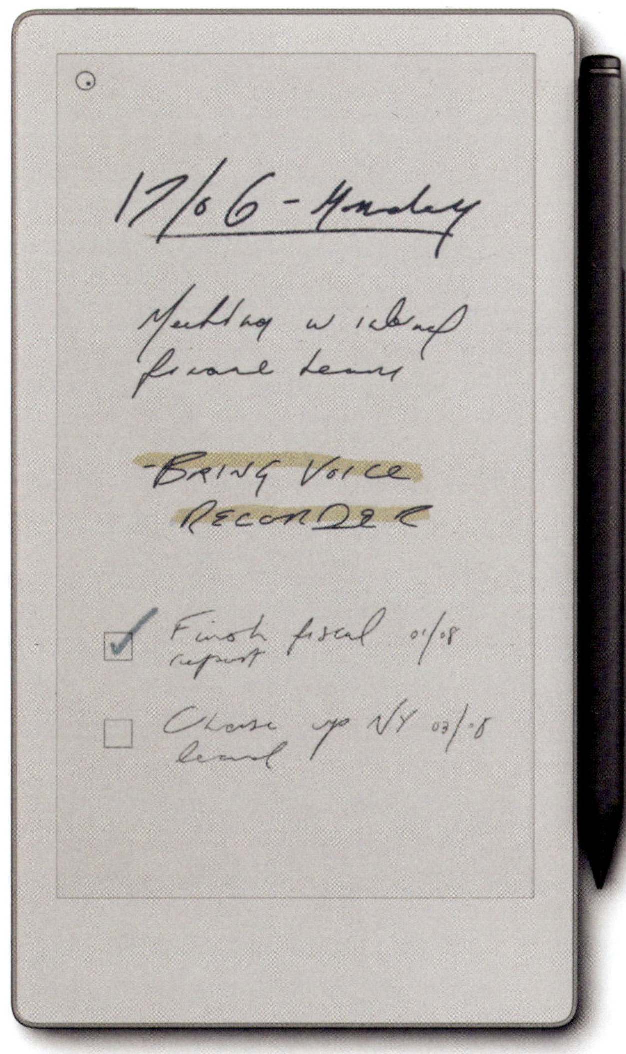

▲ *reMarkable Paper Pro Move (2025)*

---

* A type of AI inspired by the human brain that enables computers to learn from data and recognise patterns, rather than relying on specific programming rules.

Jonathan Yeo

Widely recognised as the preeminent portraitist of his generation, Jonathan Yeo is celebrated for a signature style that blends photorealistic precision with expressive, abstract mark-making. His subjects form a definitive cultural map of the 21st century. Beyond historic commissions of King Charles III and Nobel Peace Prize laureate Malala Yousafzai, his diverse portfolio includes Hollywood icons Nicole Kidman and Dennis Hopper, artist Damien Hirst and naturalist and conservationist Sir David Attenborough. When not painting, he extends his practice by designing innovative tools, merging 3D modelling, virtual reality and sculpture to lift his artistic mark into new, three-dimensional forms.

What do you consider to be your tools these days?

On one level, the materials, especially oil paints, have only evolved very slowly and marginally over the last few centuries. There were other kinds of paints and other ways of making pictures, but nothing really that's improved in terms of the range of illusions you can create. On the other hand, you could argue that the sort of materials involved in my job stretch as far as the means of travel to get to the people or the places I want to paint — and to bring my studio equipment there. Certainly, photography has meant that you can create visual notes so much more quickly than you could in the past, when you had to rely on sketching things by hand. In my case, for doing a portrait, I can capture people in lots of different angles, which means I can see them in their movements and different expressions without them having to be here all the time. And then using digital tools, Photoshop and other image manipulation software, means you can experiment or distort things by collaging the photos as a way of creating a sort of template for how a painting is going to look. Sometimes, also using the iPad and the Pencil, which is a nice way because you're still working by hand in some sense, but the range of things you can do and the ability to undo things is obviously much easier digitally than it is on a canvas.

How has your curiosity in new tools influenced your art?

In some cases, it's made things possible that wouldn't have been otherwise, such as creating portrait sculpture with a combination of 3D scanning and 3D drawing. In other ways, it's really the fact that it has allowed a certain kind of quick experimentation that would take much longer on canvas and involve taking risks of jeopardising what you've done already when it seems to be going well. What I often do when I'm halfway through a painting and have an idea about how to change it is: I'll take a picture of it, put it on the iPad, and do all the changes there. Technically, I could have done a separate picture and experimented on that or taken a photo and printed it — as I've done in the past — but it would have been much slower and much less flexible. Those are two obvious ways in which it has influenced my art.

Why did you start using 3D creative tools?

It was slightly accidental. I'd always wanted to make sculpture and had always been interested in the idea of 3D design. I didn't have any idea how to get started, and I had rather assumed it would be a long and torturous learning curve, whichever way I went into it. Then Google got in touch, wanting to demo and have me give feedback on Tilt Brush, which was their 3D drawing software. I found it very easy to use and very natural, although they didn't see it as a tool for making real-world objects. It was entirely about making virtual sorts of pictures, and that was an issue for me. At the same time, I had been introduced to this company called OTOY in Los Angeles who makes Octane — a rendering engine, but also has a light stage scanner to do sci-fi films and other kinds of VFX projects. It happens that it was being run by some people who had been art students themselves, and they liked the idea of having artists make use of their equipment to see if they would come up with something interesting. There was a day when I was in LA, and it wasn't being used. So we went in and did this very detailed capture of my face. It was just that coincidence that I had been learning 3D drawing in parallel. But in Tilt Brush, you couldn't see anything except an empty game environment. And then we found we could bring the 3D model of my face there, and I could look at it like being in a virtual studio and having myself as a model!

▼ *Jonathan Yeo's Studio with 3D Painted Portrait Sculpture, Cast in Bronze (2017–2018)*

How did it inspire you to build your own tools?

Not knowing much about technology was a sort of blessing and a curse, but we worked out a way of making my 3D portrait into a sculpture cast in bronze. It was exhibited at the Royal Academy in 2018 and got a lot of interest because no one had worked like that before, so quite a few people got in touch about whether they could commission me to get involved in doing more sculpture. One project in particular that really interested me was this idea of doing a big public sculpture of Alan Turing. One thing I hadn't managed to do with the first piece is that I wanted to get a sense of the mixture of the realist and the abstract, which is something you see in a lot of my painting works, where some bits are recognisable, some are more suggested and some are completely left out — or unrecognisable. I think there's something really interesting in the connection with how we see, experience and remember the world. Another thing that I had tried in the past was making some anamorphosis — where you break an image up into lots of different sections and then put them in different planes, either hanging them, or putting them in a Perspex block — but the interesting thing then was: could you fuse the two together and have something which was also sculptural? I realised that the process of making it so that you preserve the accuracy of the image from one angle and make it interesting looking from another — and also have all the pieces connected — wouldn't be straightforward. That's when we figured out a way of creating a bit of software that would allow you to draw the thing freehand, like we'd been doing with Tilt Brush, and then move to another point of view, locked in the perspective of the viewer so that every mark that you'd made wouldn't interfere with the image you saw from that angle. Suddenly, you were able to completely construct any sort of arrangement of the pieces, knowing it would still always look like the subject of the picture. Obviously, it took quite a bit of time to fine-tune, and the next stage we're working on now is with the foundry, Pangolin, who are well-established, brilliant and seem to enjoy doing unusual projects!

Is fun an important property of the tool?

Yes, the kind of enjoyment of using a tool, whether it's the physical enjoyment to make marks or exerting control if it's a musical instrument, is an important part of it. Not one you're necessarily consciously thinking about, but it's obviously going to be part of what makes you want to carry on using it. And ultimately, the more you want to use a tool, the better you're going to get at using that tool and probably the better you're going to be at creating whatever art form you're trying to create. Maybe it's the most important aspect of a tool that it should feel enjoyable to use. When you're doing broad border brush strokes, sometimes they're wrong, but they're faster to make. So you feel you haven't lost so much, and sometimes you find a way of suggesting multiple things with one mark that actually is more interesting than if you had explicitly shown what you were meaning by going into a lot of detail. I think that's part of the fun, and that's partly why we enjoy the works by artists who use big expressive marks rather than the photorealist illustrative image-making.

Were porn magazines fun to use as a creative tool?

Well, partly it was a practical thing. I was trying to make things which looked a bit like broken up paintings, like I do when in the early stages of a painting — a slightly cubist version — and they looked a bit like collages. I thought 'well, I'll make some collages and then, if I'm going to be making pictures of people, then it'd be quite funny to use porn mags' — and it's logical,

▲ *Bush by Jonathan Yeo, Collage on Board (2007)*

because it's skin tones! But actually, it turned out not to be such fun, in the sense that it was much harder than I realised, because you can't just make the thing directly. You have to plan it out as a painting, then translate it and take quite a long time to find the right bits of photos in the magazines that are big enough, in the right colours and have a little bit of detail that lets you know they've come from porn mags. So it was harder work than expected. But the fun came in seeing people's reactions to the toolkit you'd used because, when people would look at them, they'd react one way and then, when they realised what the medium was, they'd change their view of it. And sometimes they'd find it funny, or occasionally they'd find it shocking or make other associations with it. So there was a lot of fun in the planning of it and in seeing people's reactions to it, but definitely, the actual process of doing it was quite hard work and less fun than making a painting!

In the age of all-capable AI, is it important that tools remain somehow limited?

Being a painter, you're very aware of how complex tools are, even the ones that seem simple. Sometimes you have the luxury of having every colour, every brush and any surface you want. But sometimes you find yourself in a position where you've only got one or two things there on that day, for whatever reason. Sometimes improvising forces you to adjust your normal way of working, and in doing so, you find a better way and that feeds into your practice more generally.

Something that occurred to me early on when we were playing around with Midjourney,* is that you'd put keywords in, and whatever you'd get out wouldn't be what you had been expecting at all. It'll be much more useful when you can really control the output — describe what you want to see and get that. And I guess the logical extension is eventually being able to think of an image and have it made for you without any physical intervention at all. But in some ways, it's the opposite of what you want! Actually, what I always find is that I start with an idea for a painting and it often turns out a bit like that — sometimes not at all — but it's definitely never exactly what I was thinking about at the start. Certainly, what you get better at over the years, when things are starting to not go the way you wanted, is working out when to correct that and when to let it happen because it's better than what you would have done otherwise. It's a bit of a mantra you hear from artists: the *happy accident*. Part of the danger with AI, therefore, is 'where's that going to come from?' other than a bit of weird hallucination because it's jumbling up the things it has seen before in the wrong way. If we take AI as the ultimate thing that does everything and makes everyone, including the artist, too lazy, then where do we get inspiration if we don't spend time doing things or trying to achieve something, going over the limitations of the tool?

Beyond the tools, what effect do you think AI will have on Art?

It's so hard to know what will happen with AI because we're just at the beginning of it. But certainly on one level, what's interesting about art is conveying things in a non-literal way. The more precise AI gets and the more believably photographic it gets, it'll just make us trust photography generally a bit less, which is a shame because it's been such an amazing medium in the 20th century. But that in itself might be interesting because it might mean that there's actually more appetite for seeing how artists do things in ways that are not literal and real. What's hard to anticipate at this stage is how we'll react to that — then — and how that'll change the sorts of image we want to see and what we use them for.

Do artists fear that people will undervalue their talent if they understand the tools behind the work?

In some cases, yes. A few years ago, I was using grids a lot in paintings because that's actually the easiest way of scaling up drawings, or however you move an image onto a bigger canvas. I thought it would save time, but found out it was making me feel I should stick to the initial plan, whereas I normally change it as I go. But the funny thing was that I'd be leaving some of the canvas unfinished — like I often do — and some artists would sometimes say to me, 'but you've left the grid, you're giving away the secret of how we work to the world!' I think they just felt embarrassed that people would see that we were using some trick to achieve something. But it's just yet another method of working. Some people hate to admit they use cameras, take pictures, as well as doing things from life or from their imagination, but a lot of artists are very open about it. I didn't go to art school and do things in the traditional way or wasn't told the right and wrong way of doing things. I've figured it out for myself over the years, so I've wasted a lot of time trying things that didn't work. Therefore, when I find something that does work, I stick to it, and I don't mind sharing that information. It often happens with AI these days.

---

* Midjourney is a generative AI program that creates images from natural language, prompts, similar to OpenAI's DALL-E.

You ask people 'That's a really interesting sort of video, what AI engine did you use?' and people don't want to tell you because probably there wasn't much effort involved. It's just a lot of experimentation — which is fair enough — to get to the point where they found something that worked, and they don't want to give that away and let someone else get there without having to spend however many months it took to figure it out!

Is it why art critics and journalists got so upset at David Hockney for writing about the tools of the Old Masters?

The funny thing there is: It wasn't about contemporary art, where the critics all disagree anyway. It was about Old Masters! And all art critics, whether they are fans of contemporary art or not, no one ever questions the Old Masters! So this idea that they might have been taking shortcuts, even though Hockney's point was actually 'look how clever and advanced they were for figuring out optics before the invention of the camera'... but obviously, he was certainly making some assumptions based on what he could see rather than any total certainty. The reaction to it? There was so much negativity. Because it was somehow so wrapped up in the minds of the critics — and some of the audience — that these artists had to have been born geniuses and just do all this stuff without any help at all. Um, it's a funny thing.

What could be an interesting future for painting, say 50 years from now?

Almost certainly, a painter in 50 years will be reacting to whatever has happened in AI because AI will go through so many stages of evolution. It may be that the whole thing accelerates because what it is good at doing is replicating in variations. And so, it would suit me if I could do one painting in a particular style and then have some AI create a whole exhibition of other images without me having to do all the other ones. And so you could constantly be evolving without having to stop every time you do something good and new, and do enough of them that they can go up into one gallery or many galleries, so everyone sees them.

I can also imagine ways of combining technology and traditional tools. What's frustrating for me is that I can think of an image, and I can doodle it quickly on paper. If I've got time, I can construct it in a painting very slowly. If I have an idea for a three-dimensional image, I could possibly find some sort of 3D modelling kit, or even clay, to make it really, really slowly. I love the idea of making furniture, so I'm always thinking of things like, 'Oh, I can make a better tool for that or a better chair or a better whatever.' But you don't want to make it so easy for people that they can tell an AI or a software to come up with something vaguely like what they're talking about. To actually be able to make the thing you're imagining, I think that would be really exciting. So, if the tools could not completely bypass the act of physicality, because that's part of the reason you learn how images and the world are constructed — by trying to replicate it by hand, but allow you to do more things than we can currently do, easily and quickly, I think that's exciting. I'd hope that — maybe ten years is a bit optimistic — but in 50 years' time, certainly there should be some great ways of creating visual images and even more three-dimensional constructions, buildings and landscapes and all those kinds of things.

Natalia Ilyin

Natalia Ilyin is a Professor of Design, Design History, and Criticism at Cornish College of the Arts in Seattle. A 2024 Fulbright Specialist grant recipient and three-time winner of Cornish's Award for Teaching Excellence, she is founding faculty of the MFA in Graphic Design at Vermont College of Fine Arts. Formerly national director of programmes for the American Institute of Graphic Arts, Natalia has taught at Yale, RISD, and Cooper Union. She bridges theory and practice by leading workshops for global organisations, including Microsoft and Boeing.

In 2019, she authored Writing for the Design Mind,[8] a vital creative tool crafted specifically for the designer. Rather than teaching rote grammar, Ilyin offers a cognitive framework that enables visual thinkers to master the written word using the very same methodologies they apply to design. By translating abstract concepts into practical, visual exercises, the book functions as an essential mental apparatus, fostering the confidence needed to excel in industry and academia while making the process of writing feel more human.

What tools play a part in your creative process as a writer?

My teapot! I have a whole routine of sitting down with a cup of tea to write. That sounds ridiculous, but it's actually really important because you have to create a construct within which to feel safe in order to get out things that are true to you. Things you really believe are hard to say sometimes. One of the things with writing is that you have to put yourself in a tunnel of focus in which you are completely free. It's a funny juxtaposition of tunnelling in and then feeling you have no boundaries and no constrictions. I guess that tool is having things that put me in a hypnotic state. One of those things is having background noise. You wouldn't think that would be conducive to focus, but I'll often put on a video in the background of people talking quietly. I don't like coffee shops because I know a lot of people, and they come barrelling over to introduce themselves. It's irritating [laughs]. I like having the sound, especially of male voices. There's something in the sound of male voices speaking quietly — that's a tool I use to focus in because I feel safe with male voices around. That's the real truth. Voices are calming. So I do that too. Kind of a weird tool.

Do you think tools for the mind and tools for the hand are fundamentally different?

No, I think they're the same thing. That's probably why I'm so in love with diagrams. There's a very difficult book that is very good called The *Culture of Diagram* by a couple of Stanford professors.[9] It's about how a diagram is the meeting of description by hand and description by language. So it's the combination.

I think they're two sides of the same consciousness, really; there's not any real difference. But then I have a design mind, and I also have a writing mind, so I'm probably not the person to ask; it's pretty integrated in me.

Reflecting on your role as an educator, you wrote an article titled 'I Am Not a Cruise Director'.[10] Do you think designers should stop trying to be 'cruise directors'?

The definition of the *cruise director* is a person who orchestrates excitement and entertainment, rather than the role of the teacher or the designer as an instructor or person who pushes people into places they may not want to go. I don't know if I've ever known a graphic designer who pushed people into places they didn't want to go. And yet that's not true. There are certain social activist designers who make things that make us feel that things are not right, and we want to make them right. Instead of being cruise directors, which points us directly at capitalism, a little bit more of the instructor point of view is not a bad thing.

The great levelling of postmodernism, where every voice was considered equal, made us feel that maybe we shouldn't take the role of the authority or the person who knows more than another person. That's the elitist conflict that has been going on. I think the ethical stance is this: if you know something and somebody else doesn't know it, and they ask you for it, you should give it to them. We have to set the stage for designers to be instructors. We do have a great opportunity to show people things they don't know and push them in directions they don't want to think. You can't go too far with that, or you become preachy and moralistic. It

▼ *Blog Post by Natalia Ilyin (2015)*

NATALIA ILYIN

ABOUT WRITING **NATALIA'S BLOG** STUDENTS BOOKSHOP MISCELLANY CONTACT

TAGS

Semiotics
Design Writing
Design for Social Change
Omnicriticism
Island Life

I AM NOT A CRUISE DIRECTOR

November 8, 2015

There's a fashionable attitude in education right now that assumes that making things more interactive and more entertaining and more experiential for students results in a better education for that student. I think we'll live along into a time in which we will look back on this strategy and see its shortcomings clearly.

takes a fine hand to push people enough without telling them what they should be thinking.

An educated designer has a bigger worldview than just making fun things. That drives me crazy. People don't spend time figuring out, 'I've got two feet on this planet here. I have to include every living thing. I can't just include myself.' I don't know how that tunnel vision is schooled into designers. It's not the people's fault; it's the schooling. It's maximum consumerism and capitalism. It's tunnel vision, and it's ridiculous. We need to get some people in there and push for a larger view.

Should designers be more critical?

Let's define critical. Critical does not mean being negative. To me, a critical designer is a person who has a larger worldview and has educated themself to understand their role in the larger society and the future of that society. Designers have a real ability to change the course of society in a way that is ridiculously large for people who are as uneducated as we are. My school has made it a point to have three classes in design history and criticism instead of one. Most graphic design students in normal universities take one class in history. That's 15 weeks of contextual information, perhaps for their whole career. That's nothing. I have students who don't know that Poland was once occupied by the Soviet Union. I also have students who are highly educated, but many just don't know much about the world.

What I actually teach when I'm teaching design history is the history of the contemporary world through the lens of made objects. Many students want more context. Educators generally don't give them much because they don't have many credits available to do so. They're always trying to mush so much into the curriculum. But to be a real designer, I think, you must always be thinking about the effects of what you do on other people and on the planet. And as a critic, keeping that question firmly in mind — that's the larger kind of criticism to me.

Did you intend your book as a tool to help designers be more critical?

Frankly, I had not intended to write a book about writing. I was asked if I'd like to write one, and had to refocus my thinking a bit to do so. But now that you say that, I guess that was in my mind. I wanted students to know that their voices are absolutely important, and the tools needed to write are no more difficult to master than any of the other myriad tools they've mastered in their lives.

You've written that the ubiquity of the computer has pushed people with certain skills out of the design industry or forced them to adapt, often with negative effects. Do you think it might have made creation more accessible to others?

No, I don't think so. I am old, and I remember the kinds of people who were attracted to design before computers. These people had infinite backgrounds. They may have grown up with a father who was a bus driver and a mother who was working in schools, and they just found their way. They saw what they liked and found their way toward it. That's different from having a paint program on your computer when you're six and thinking that that program formed the parameters of creativity. Computer applications created stereotypical pathways for the mind and helped us think we were creating. Working with apps is creating, but creating according to very specific standards and rules. But real, free creation is accessible to everyone as soon as they start thinking, 'I want to make something'. I don't think computing has made creation more accessible; I think it's made it more regimented.

▲ *Writing for the Design Mind by Natalia Ilyin Published by Bloomsbury Visual Arts (2019)*

Do you think software tools have a natural tendency to lead in a direction of their own, whereas a pencil is more likely to follow the artist's intent?

David Berreby has a Substack called *Robots for the Rest of Us*.[11] He was just saying yesterday that he felt he was becoming a part of AI rather than AI becoming a part of his life. At the very beginning — around 1989 — computers had a humongous tendency to lead somewhere of their own intent. All the decisions made when you were in Photoshop had already been programmed in for you. The people who created those early applications had their own taste, their own viewpoints about what design was and what it should do. Adobe's designers made a tremendous template, using their own views. When you open up InDesign, you have options about what size to make the page. It doesn't give you the option to make a round page; you have to work to make that happen. No, a nice, normal page is what's presented to you first.

What do you think motivates people to use certain tools?

I have a snap answer to that: they use what the people sitting next to them are using. I originally picked up a paintbrush because my father was a painter and there were brushes all over the place. I picked up making garments because my mother made garments. I did it because somebody near me was doing it.

To your point, creation often begins with imitation. Yet, with software — for instance, when we follow a YouTube tutorial — imitation does not allow for anything to get lost in translation; the same tools yield identical results. What effect, then, does this have on creativity?

Can we talk in semiotic-speak for a moment because it's so appropriate? The sign is transmitted when I pick up the paintbrush my father used. The sign (paintbrush = creativity) is transmitted through a kind of filter of many cultural impressions, personal impressions. The paintbrush, as the idea of painting or what I can paint, is coming through a mental filter of incredible complexity; I don't recognise it, and the sign comes into my mind loaded with personal and cultural implications. That unconscious loading makes a new sign in the mind of the receiver, called the *interpretant*. The *interpretant* is severely reduced in the direct imitation from YouTube. It doesn't have any personal nooks and crannies. The paintbrush means all those things that have come through the filter to me, and the final result is that, with apps or YouTube videos, my experience of creativity is reduced, though I do not realise it.

With a computer, when you're going to copy a mountain in 3D with a tutorial: you get the thing, you look at the thing, you do the thing. Your filter is almost empty of experience or cultural information, aside from that one little YouTube cultural knowledge base. It is almost empty, so that the sign of the digital paintbrush enters my mind as far more exactly transmitted, instead of being refracted against my experience and that of my culture. So it is a hollowed-out kind of experience or a kind of signification that I end up transmitting because I haven't had many different experiences to attach to it.

Is AI an even greater threat to the tools of the mind than it is to physical ones, simply because the immaterial is harder to see?

Yes, it's a bigger threat — but only if you consider individual creation important. It depends on what the society values. It all comes back to how the people who have the power right now want to order the society. You and I look at a reduction of personal creativity options as a threat to individual response, response formed by our own personal experience, our own identity. Techbros don't look at us that way. They give people the impression that they're writing when they're not writing, keeping them busy so that they don't get out in the streets and revolt. Bread and circus, as you say. If Minecraft had not been invented, the revolution would be over by now.

Because AI is invisible in the creative fields, it is a much bigger threat to the culture because it hollows out the authentic experience of the people of the culture and replaces it with a sort of soft persuasion that all is okay and we're all just snapping to grids here, and everything is clean and nice as we tap along unconsciously.

If we lose these immaterial tools, are we at risk of losing part of our identity as well?

First of all, as soon as you read something that you're really paying attention to, it becomes a part of your materiality. There is no such thing as an immaterial tool; every experience of a tool is framed in your neural networks. With AI, many people will lose immaterial tools, and they will lose part of their identity, and they will not know it. The risk of loss of identity — we have to step back and say, 'What is identity today?' 'Have we already thrown in the towel?' 'Have we already been usurped?' I think most designers have already lost part of their identity, and the battle here is to regain it.

It is often easier to spot the Orwellian (mass surveillance) than the 'Huxleyan' (mass distraction). Are we perhaps mistaking entertainment for creation?

Oh yes, oh yes, oh yes. Creation is not entertainment or entertaining. It's painful. Real creation — ever tried to birth a baby? It's the same with writing — for instance, the birthing process you are going through right now with this book. Creation is unpleasant because you're stepping out of the known, and you are making a bet that the unknown can be known. That sounds insane, but stepping out is rather like Wile E. Coyote stepping over the cliff and not yet realising he's in mid-air. That is the feeling of real creation. It is unpleasant and difficult, not knowing what you're doing.

Remembering the teapot — that's why writers and artists (and writers are artists) have to create safe spaces for themselves. Real making is hard. You're constantly fighting with: 'Am I going crazy?' or 'Am I on the wrong track?' All of this back-and-forth with yourself is very difficult. But there's none of this soul-ripping when you can use InDesign and just make something look clean and pretty, quickly and easily. And that's a good thing. There's a kind of design that is fast and efficient, and we are not talking about that. You don't want to rip your guts out over every hangtag you design. You don't want to fall on your sword about every cheese package you design. So there's stuff the mainstream can just do and take care of quickly, and that's a good use of the current applications. But if you want to really make something, I'd say abandon the applications and figure it out on your own. Techbros are keeping a lot of people entertained with making things they think are design, but are actually rule-following within strict limits. But you know? There are a lot of designers in the world, and there isn't that much room for truly unusual and different designers.

If you took a more 'Huxleyan' approach, you could say it's actually good that many people are starting to mistake entertainment for creation; it keeps them out of the way while those who do not make that mistake can create their work and be lauded for it. If you hold that point of view about creation, it's cynical — but it's true.

Current AI tools offer instant results and fast iterations in lieu of a journey. Can AI be considered a creative tool if the creative journey is missing?

No [laughs]. I'll be honest: I do not write with AI; I even have a disclaimer on everything I write that says, 'this wasn't written by AI'. My students do that too. I do use AI to make sure that I'm quoting something correctly or to add a citation. For me, it's a research tool. I never trust it, but it'll aim me in directions that I can double-check — I use it for quick listings mostly. In that way, it is a creative tool because it allows me to come up with things quickly that I then make decisions about faster. I can use it to do research faster — even if I'm not using it as final research — winnowing things down and seeing if an idea has enough heft behind it.

I went off on the word *passion* a few years ago, because everybody was saying how *passionate* they were about graphic design. Passion is not a happy thing; a passion is a moment of pain — passion *means* pain. We have to really think about the word creative too. *Creativity* does not mean *art*, and it doesn't mean making things pretty. It doesn't mean having fun either. It doesn't mean any of those things. It means to focus on a problem, a story, or something that irks you, and apply your one and only personal cluster of neural networks to spark a response.

For AI — and I'm talking about large language models — to be valuable to the creative journey, it has to be used to back up this true creativity and not endeavour to replace it.

When you hear someone say 'it's just a tool,' what is it you'd like them to consider?

I want them to think about the word *just*. What does *just* mean to them? What are we downplaying when we say it's *just* a tool? What are we assuming about tools? That they don't have agency over us? Maybe I should say, 'That they don't have agency over us — or do they?' like ChatGPT.

---

[1] Albert Einstein, quoted in Ronald W. Clark, Einstein: *The Life and Times* (New York: World Publishing Company, 1971).

[2] Garson O'Toole, 'Quote Origin: My Customers Would Have Asked For a Faster Horse', *Quote Investigator*, July 28, 2011, quoteinvestigator.com/2011/07/28/ford-faster-horse/.

[3] Henry Ford and Samuel Crowther, *My Life and Work* (Garden City, NY: Garden City Publishing Company, 1922).

[4] Malcolm Gladwell, *Outliers: The Story of Success* (New York: Little, Brown and Company, 2008).

[5] Arthur C. Clarke, 'Hazards of Prophecy: The Failure of Imagination', in *Profiles of the Future: An Inquiry into the Limits of the Possible* (1962; repr., New York: Harper & Row, 1973).

[6] Run The Joules, 'This reMarkable 2 tablet does (almost) nothing...', YouTube video, 19:54, April 2, 2023, www.youtube.com/watch?v=yLbAh2oBPUO.

[7] Alessandro Scarano, 'reMarkable 2, the device created as an improvement to paper', *Domus*, 10 August 2021, domusweb.it/en/speciali/modern-work/gallery/2021/remarkable-2-the-device-created-as-an-improvement-to-paper.html.

[8] Natalia Ilyin, *Writing for the Design Mind* (London: Bloomsbury Visual Arts, 2019).

[9] John Bender and Michael Marrinan, *The Culture of Diagram* (Stanford: Stanford University Press, 2010).

[10] Natalia Ilyin, 'I Am Not a Cruise Director', *Natalia Ilyin* (blog), November 8, 2015, nataliailyin.net/blog/2015/11/8/i-am-not-a-cruise-director.

[11] David Berreby, *Robots for the Rest of Us*, Substack.

3 | Iconic Tools

We usually speak of iconic designs for objects that have marked a period in time, whether by being in everyone's hands, like the Coca-Cola bottle, or confined to our collective imagination, like the Ferrari F40. Of course, some designs are only understood to be iconic by a handful of like-minded individuals, which can make for great heists. Infamously, at the turn of the millennium, hundreds of Pierre Jeanneret's V-Leg chairs, designed in the 1950s for government offices in Chandigarh, India, could be found in junkyards until collectors started flocking in. Today, they often sell for more than £10,000 apiece.[1]

For creative tools, iconic status does not come from dominating the *consumer* imagination or from a hefty price tag in design galleries. Because their existence as *objects* of design is secondary to their *ability* to enable human creation, their iconic status can only be established through that lens.

Hence, all the designs in this chapter have been chosen for their direct and measurable impact on creation. Typically, they may have been used for famous works of art by renowned artists, or they may have made creation accessible to a much wider audience than was previously the case. Sometimes, they may even have been a fork in the road that influenced all subsequent designs or led to the birth of entirely new artistic movements and significant cultural shifts.

With such criteria, it appears that few of the names traditionally associated with design have produced creative tools that could be considered even remotely iconic. Marcello Nizzoli's Olivetti Lettera 22 typewriter, Luigi Colani's Canon T90, and Sir Jony Ive's Apple iMac and iPad are the exceptions rather than the rule.

1 | Ochre & Charcoal

ca. 70,000 BCE — Pigments like ochre and charcoal mixed with sap, egg yolk, wax, oil, fat, or resin are used for symbolic and artistic painting.

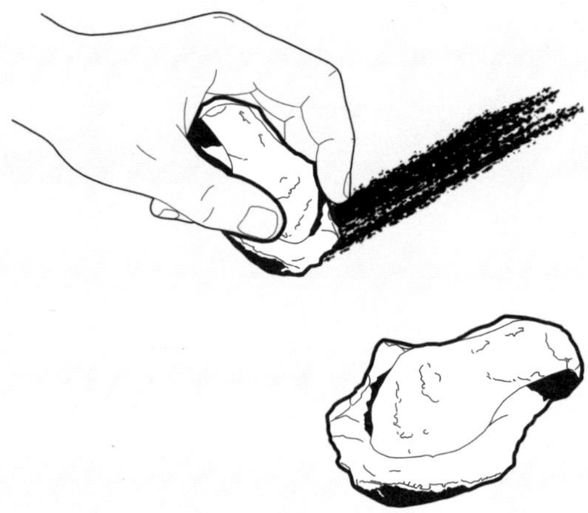

2 | Bone / Ivory Flute

ca. 60,000 BCE — The earliest known musical instruments made from animal bones provide clear evidence of prehistoric music-making.

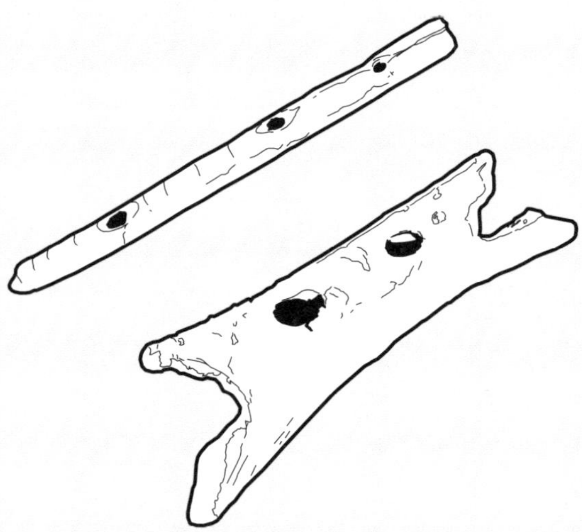

3 | Bone Airbrush

ca. 40,000 BCE — The first mouth-operated airbrush made from bird bones is used to spray iron oxide pigment to create cave paintings.

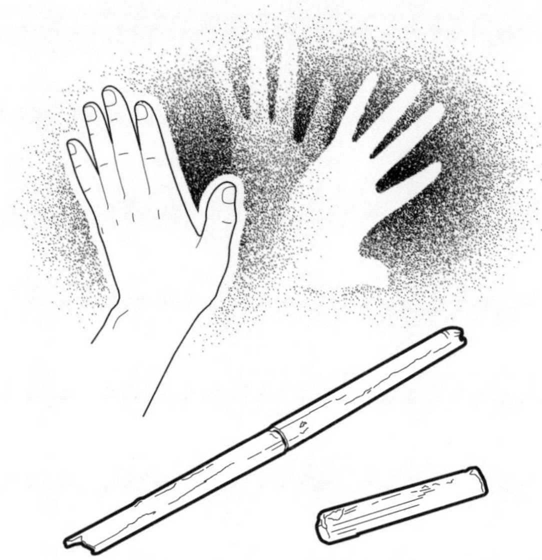

4 | Bone Needle

ca. 20,000 BCE — The origin of textile arts and tailoring, with a design form that remains recognisable and effective even today.

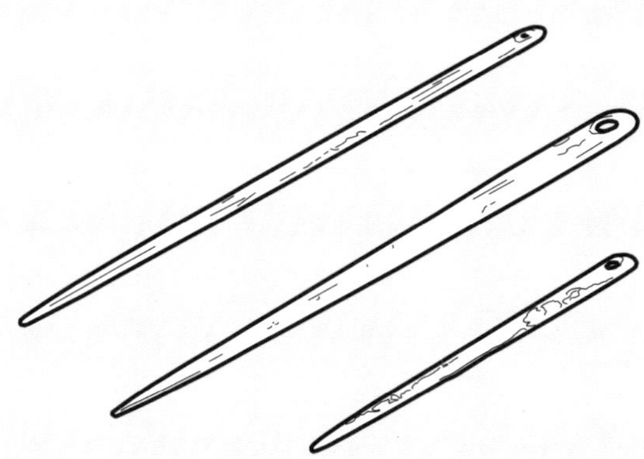

5 | The Lyre

ca. 4500 BCE — An instrument from Greek mythology, the lyre is said to have been gifted by Hermes to appease Apollo, a god of music, song, dance, and poetry. With a history spanning ancient cultures, including the Egyptians, Greeks, and Germanic tribes, its origins can be traced back to the ancient Middle East.

One of the oldest surviving examples of any stringed instrument, the Sumerian Queen's Lyre from the Royal Cemetery at Ur is made of wood, with intricate inlays of precious materials and a resonating soundbox. Its gut strings are strummed or plucked to produce a clear, bell-like tone.

► *The Queen's Lyre (ca. 2500 BCE) Partial Reconstruction*

6 | Potter's Wheel

ca. 3500 BCE — A machine for shaping clay into round ceramic vessels, the earliest designs were slow-wheel turntables, moved by hand to increase the speed of coiling a pot.

The fast wheel, using a heavy stone flywheel and operated by stick or foot (a kick-wheel), spun continuously and enabled the *throwing* of clay. This design enables faster production, greater symmetry, and thinner, more refined vessels that support growing urban societies' need for standardised vessels for storage, cooking, and trade.

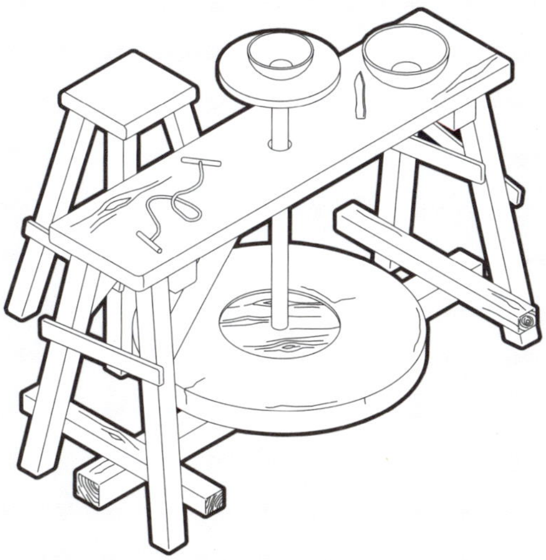

7 | Writing

Category: Writing	
Release: ca. 3200 BCE	
Designers: Sumerians, Egyptians, Chinese and the Olmec People in Mesoamerica	
Manufacturer: The human hand	
Dimensions: A standard ruled line on a modern paper page is around 8mm high	

▲ *Clay Cuneiform Tablets, Assyrian (ca. 20th–19th century BCE)*

Emerging ca. 3200 BCE in the earliest civilisations, such as Sumer in ancient Mesopotamia (now south-central Iraq) and Egypt, writing was born of practical need. As trade and governance outgrew human memory, early designers of writing systems created marks to track grain, livestock, labour and value exchange. In doing so, they externalised working memory, recording and manipulating ideas too large to fit in the conscious mind.

Writing shifted gradually from accounting to creativity as it became easier and less permanent, inviting reflection and revision; exploration, persuasion, and imagination became possible. Writers noticed that the act generates feedback: ideas clarify, mutate, and multiply when made visible, and writing revealed itself not just as a record of thought, but as a tool for producing it.

▲ *Wooden Block Movable Type*

Design & Features

Writing satisfies three defining criteria: meaning, standardised symbols, and a correspondence to spoken language. This separates it from proto-writing, which relies on ideographic or mnemonic marks that gesture toward meaning but cannot fully encode a grammar and lexicon.

Deliberate choices about symbols, materials, and structure play a crucial role: Cuneiform wedges pressed into Mesopotamian clay emphasise durability and record-keeping; Egyptian Hieratic scripts on papyrus favour speed and fluidity; inscriptions carved in stone convey permanence and authority; and ink, paper, print, and movable type expand accessibility and aesthetic range. Each case balances ritual, beauty, practical efficiency, and stability. Stability enables durability and distance, meaning that writing can outlast its creators and travel far beyond its point of origin.

Typical Workflow

1. Clarify the intention and define the purpose, audience, and any constraints that will shape the work.
2. Decide which tool to use; the choice will affect both process and the final work: form and style, pace and rhythm, ability to revise and share, the cognitive experience, and aesthetic and cultural value.
3. Externalise thoughts into writing, prioritising flow and discovery as ideas move from internal cognition to symbolic form.
4. Reread and refine, revising to improve clarity and sharpen meaning.
5. Stabilise the work and share it as a durable artefact that can circulate, persist, and influence others.

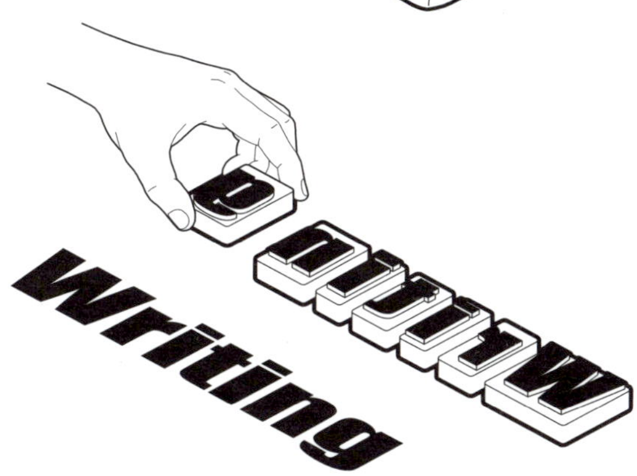

Writing Tools Cultural Impact

Writing has reshaped human societies. In emerging complex civilisations, it enabled humans to externalise memory, record economic transactions, and create a stable, lasting record of laws and agreements. In this sense, writing functions as extended cognition, complementing internal thought by making concepts visible, manipulable, and retrievable across time and distance. As writing moved to expressive and reflective uses, a profound cultural shift occurred. Literate scribes, trained in specialised schools, became powerful figures in ancient courts and temples, controlling access to knowledge and shaping bureaucratic and religious life. They drafted contracts, copied stories, and maintained archives that powered legal, economic and religious systems.

Building on oral traditions, the creation of classical narratives, such as The Epic of Gilgamesh (ca. 2100 BCE), demonstrates how writing facilitated enduring artistic expression and scholars such as Walter J. Ong have argued that the shift from orality to literacy restructures consciousness itself.[2] In oral cultures, knowledge resides in communal performance and memory wheras literacy introduces individual reflection and abstract reasoning, and writing makes detached analysis and cumulative scholarship possible.

Writing enabled the transmission of scientific, philosophical, and literary thought across generations. However, early writing systems were difficult to master and often restricted to elites, reinforcing social hierarchies and limiting broader participation. Material constraints and literacy barriers meant that for much of history, most people remained excluded from written knowledge. While intended for stability, writing became a tool for both subversion and social change, continually reshaping the possibilities of human society and culture, for good and bad.

8 | Wax Tablet & Stylus

ca. 715 BCE — A hollow wooden base filled with wax and coated in dark soot allows for writing and *erasing* with a bone, wood, or metal stylus.

10 | Quill Pen

ca. 6 CE — Crafted from the hollow wing feather of a large bird, holds ink and flexes enough to scribe letters on parchment and vellum.

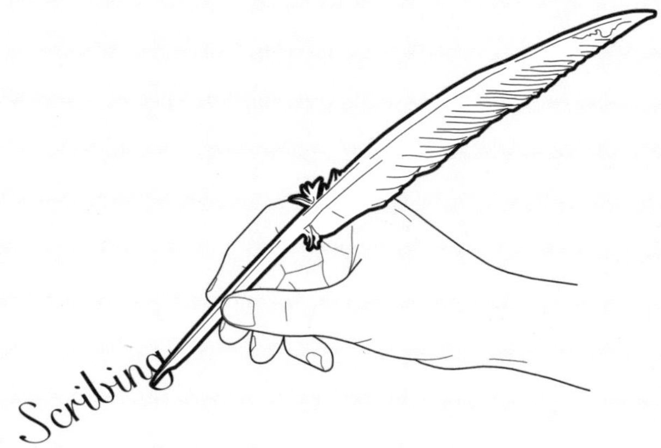

9 | Compass / Divider

ca. 79 CE — Used for scratching arcs, drawing circles, and measuring or transferring precise distances in geometry and navigation.

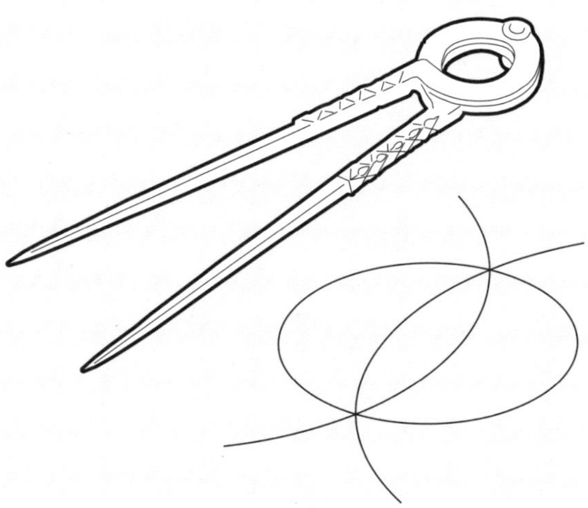

11 | Moveable Type

ca. 1040 — A configurable system for printing characters and words on paper, using stamps made of wood, clay, ceramic, and later, metal.

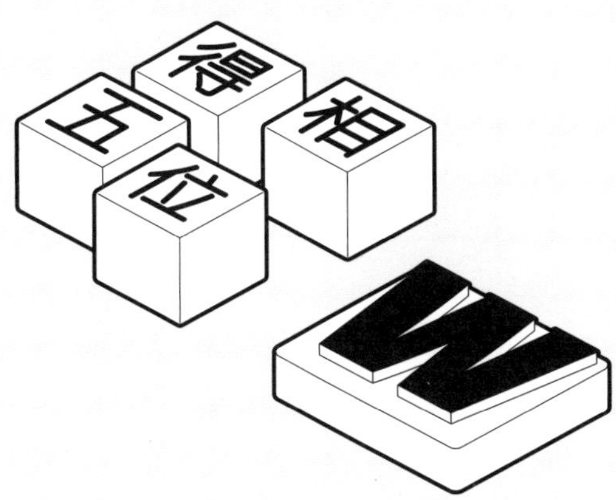

12 | Silverpoint Stylus

ca. 1400 — Fine drawing in the Middle Ages and Renaissance used silver, a soft metal that readily marks prepared paper, parchment, and wood.

13 | Gutenberg Press

ca. 1450 — German Johannes Gutenberg creates the mass printing press, setting pages with a matrix of uniform, durable metal type.

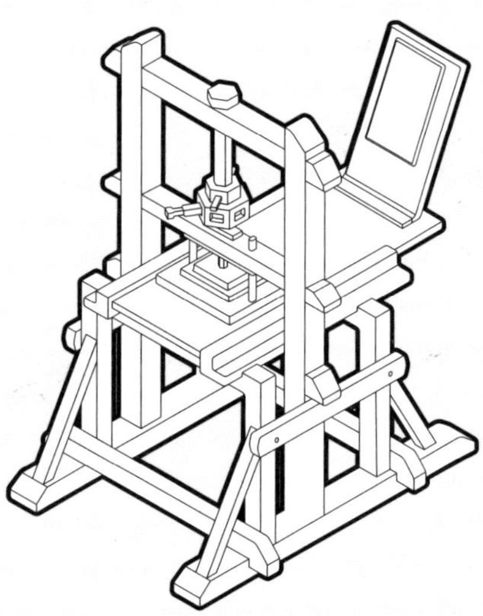

14 | The Burin / Graver

ca. 1450 — A controllable hand tool with a tempered-steel cutting tip can finely engrave metals with precise lines and stippled patterns.

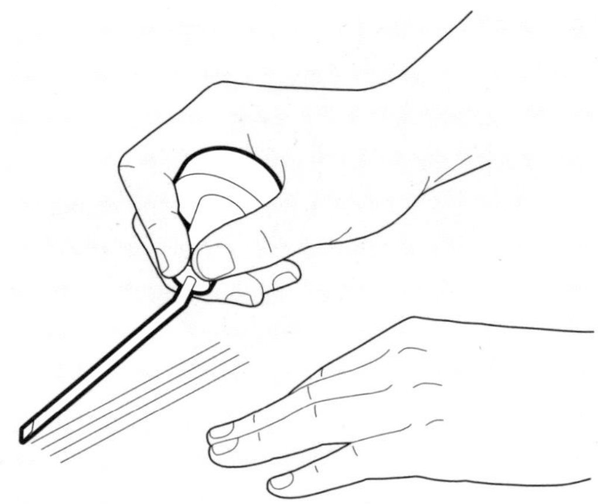

15 | Camera Obscura

ca. 1500 — An optical device that projects an image of the outside world onto a dark interior surface, using a small aperture or focusing lens.

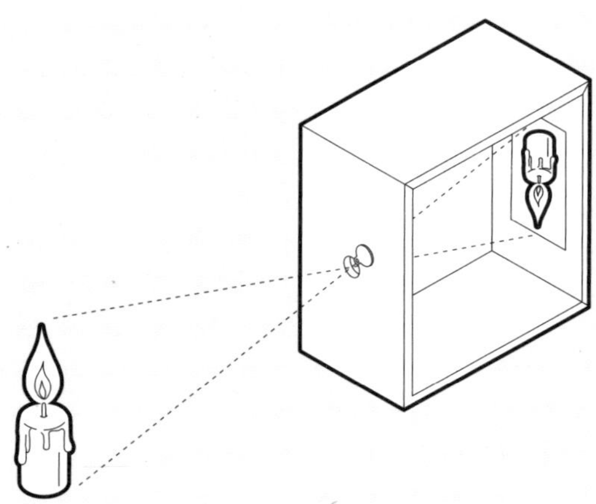

16 | Muller & Slab

ca. 1600 — The stone or glass muller was essential in artists' studios to grind pigments into oil, producing smooth, workable paint.

18 | Jacquard Machine

ca. 1804 — Attached to a loom to automatically control weaving using a sequence of punched cards that correspond to complex patterns.

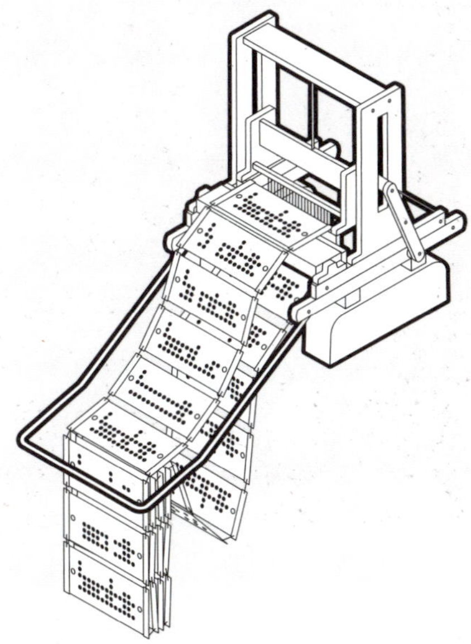

17 | Fortepiano

ca. 1700 — Invented by Bartolomeo Cristofori, the fortepiano has leather-covered hammers, thin strings, and a light wooden frame without metal bracing. Its lighter action produces a softer, more expressive sound with less sustain than that of the modern piano.

Wolfgang Amadeus Mozart praised it, writing, 'One can hear the difference immediately when one plays softly or loudly; the tone is always even.'[3] By contrast, the harpsichord (ca. 1390) plucks strings and allows little dynamic variation.

► *Piano by Bartolomeo Cristofori (1720)*

19 | Winsor & Newton Paint Tube

Category: Art, Illustration

Release: 1840

Designers: John Goffe Rand, William Winsor

Manufacturer: Winsor & Newton, United Kingdom

Dimensions: approximately 25 mm × 90 mm × 20 mm

Until the mid-19th century, artists relied on fragile methods for storing paint, grinding pigments by hand and placing them in pig bladders, jars, and glass syringes. These proved messy, short-lived, and impractical for travel. In 1841, American portrait painter John Rand patented the collapsible tin paint tube, developed and commercially sold by William Winsor: a small, flexible container that could be sealed tightly, preserving premixed paint for extended periods.

Artists were no longer bound to the studio and could carry many colours outdoors, painting en plein air, capturing motion and changing light with spontaneity and expressiveness. This new mobility and flexibility helped fuel the birth of Impressionism: Monet, Renoir, and Pissarro all painted new subjects and styles, carrying paint tubes.

Design & Features

A slim, flexible tin cylinder can be squeezed to dispense paint and sealed with a screw cap to prevent drying. The collapsible form ensures efficient use of its pre-prepared contents while keeping it airtight and stable.

A fragile material is turned into a dependable, portable tool, and the act of dispensing is now a precise, deliberate, and repeatable gesture. Standardising pigment storage, the tube allows painters to expand their colour palettes, mix with greater control, and achieve new levels of precision.

Lightweight and durable, the tube fits easily into a pocket or paint box and encourages work outdoors and whenever inspiration appears. Aluminium and plastic have replaced tin today, and the design is even lighter and more resilient, but the original essential form and benefits remain.

Typical Workflow

1. Select from a range of pre-mixed colours stored in sealed tubes. The convenience of ready-to-use pigment eliminates the need to grind and mix, allowing more time for creative exploration.
2. Gently squeeze the flexible tube to dispense a measured amount of clean, fresh paint with a consistent texture. On the palette, mix colours and experiment with tone and light.
3. Replace the screw cap to protect the remaining paint, and leave it ready to reproduce the same colour later.
4. Carry the palette from the studio into the landscape. Paint outdoors and capture fleeting moments, and the effects of weather and time. Use the paint's freshness to achieve quick, expressive brushwork that reflects direct observation.

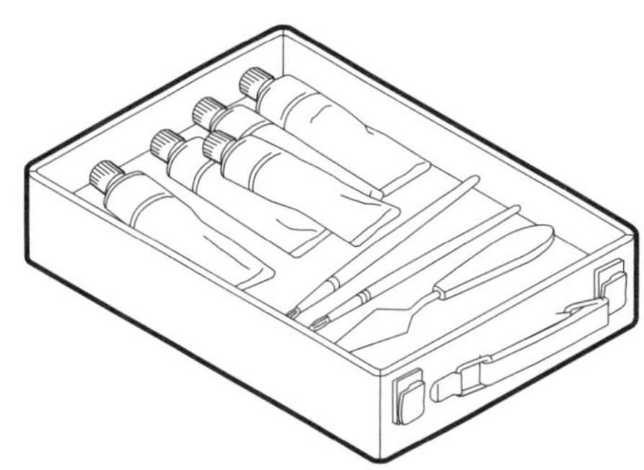

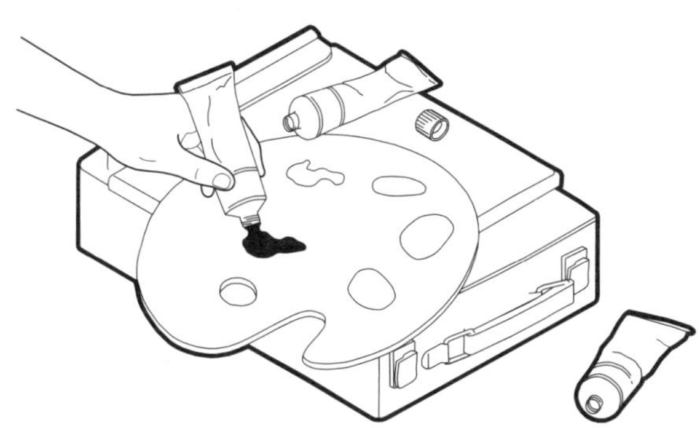

20 | Singer Model 15

Category: Making

Release: 1879, 1895 (improved), 1920 (electric)

Designers: Elias Howe, Isaac M. Singer

Manufacturer: Singer, North America

Dimensions: approximately 370 mm × 300 mm × 180 mm (without base)

The Model 15 is one of Singer's most significant achievements. In production for more than a century, it became the foundation of modern sewing practice, was mass-produced, and widely distributed, bringing a level of craftsmanship into the home that had previously belonged to industry.

Its importance lies in both its role in expanding access to sewing and its symbolic presence as a tool of creativity, independence, and self-reliance. The Model 15 enabled people to create, repair, and design garments with professional-quality results. From everyday household sewing to small-scale tailoring and fashion work, it became a cornerstone of creative domestic life and inspired generations of makers and artisans.

Design & Features

The oscillating shuttle hook produces a strong and balanced lockstitch, resists thread jams, and is easier to maintain than in earlier machines. The hand wheel, side-mounted tension assembly, and vertical needle bar are lasting design benchmarks. Later electric versions introduce a motor without altering the essential mechanical design, underscoring the purpose and durability of the original engineering.

Visually, the Model 15 conveys a sense of industrial elegance: its black enamel body and fine filigree details reflect the craftsmanship and aesthetic values of its era. From the smooth motion of the components to the gentle hum of the mechanism, the ergonomics encourage the user to feel each movement, listen to the rhythm of the stitches, and sense a balance between hand, fabric, and machine. It is a tool that rewards both skill and sensitivity.

Typical Workflow

1. Oil the moving parts, thread the needle, and adjust the tension knob (T) with care. Patience and precision come before speed.
2. Use the lifter (F) to raise the presser foot and place the fabric beneath it.
3. Set the machine in motion by starting the treadle and wheel rotation with your feet and hand, or use the foot switch to activate the electric motor, while another hand guides the fabric.
4. Follow the mechanism's steady oscillation and let the cloth move naturally. Guide it gently, neither forcing nor resisting the consistent pull of the feed.
5. Fine-tune thread tension and stitch length. Allow the machine to 'answer back' through its sound and the feel of its resistance.

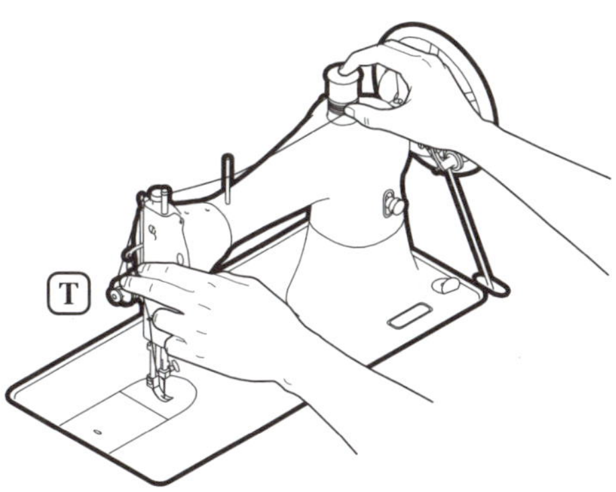

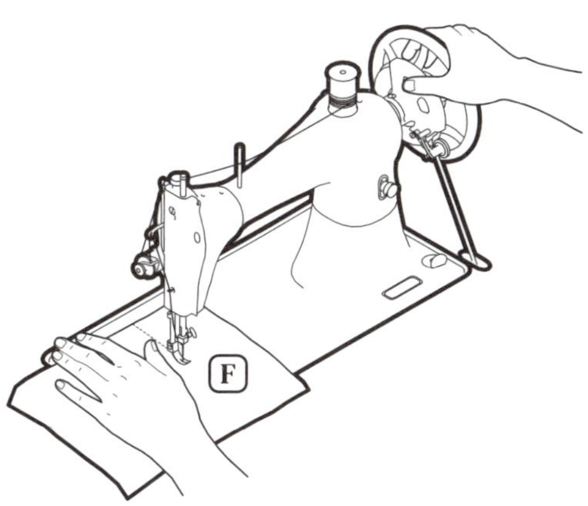

21 | Underwood No. 5

Category: Writing

Release: 1900

Designer: Franz X. Wagner

Manufacturer: Underwood Typewriter Company, North America

Dimensions: approximately 305 mm × 305 mm × 457 mm

The Underwood No. 5 is widely regarded as the first truly modern typewriter, establishing the form and functionality that would define the machine for decades. Produced from 1900 to 1931, it enjoyed an extraordinary manufacturing run of nearly four million units.

Its influence spread across offices, newsrooms, and writers' desks. Many of the 20th century's most celebrated authors, among them F. Scott Fitzgerald and Ernest Hemingway, relied on the Underwood No. 5 to craft their creative works. For more than a generation, it stood as the benchmark for personal, professional, and creative text production, uniting thoughtful design with dependable mechanics to create a tool that was both elegant and highly capable.

Design & Features

The design implements a front-strike mechanism that lets typists see their words as they appear on the page. A precisely engineered cam action within the typebar linkages accelerates each letter stamp, creating a crisp, responsive action and a consistent impact, resulting in tactile confidence.

At the end of each line, a quick-return lever smoothly returns the carriage to its starting position, enabling a fast, continuous work cycle. The standard QWERTY keyboard uses a single-shift mechanism, making it easy to toggle between lowercase and uppercase, and an early form of shift lock allows uppercase to be held without hand strain. A wide carriage accommodates long lines of text, making the machine adaptable to everything from personal correspondence to professional documents and publishing tasks.

Typical Workflow

1. Roll a sheet of paper into the carriage. Align it with the platen (rubberised cylinder) (P) using the paper release and feed controls. Check the ribbon is inked and threaded correctly, and begin typing.
2. Each character appears instantly on the page. Decisions are immediate and physical. Let thought and rhythm flow with the 'clack' of keys, developing a deliberate cadence.
3. As the carriage nears the margin with a bell sound, push the carriage return lever (R) to bring the platen back to the start, advancing the paper forward one line. This small ritual marks the pulse of composition.
4. Edits are manual and visible. Cross out words, overtype letters, and insert marks. Full revision is a process of retyping rather than editing.

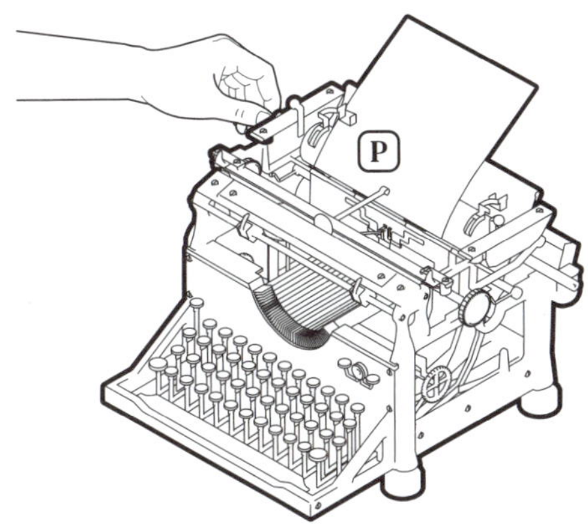

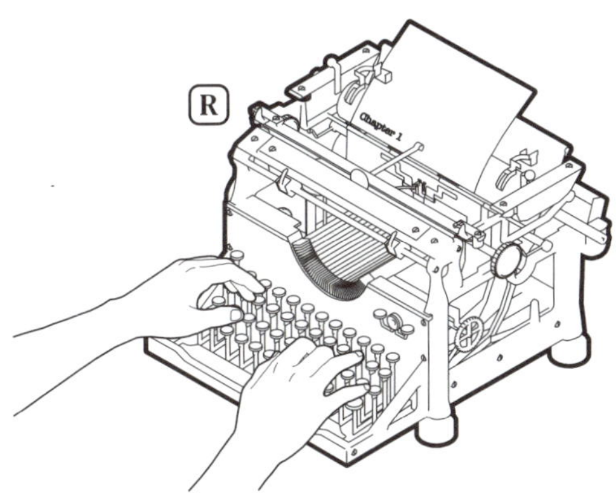

22 | Leica 1

Category: Photography

Release: 1924

Designer: Oskar Barnack

Manufacturer: Ernst Leitz Optische Werke, Germany (the Leitz factory)

Dimensions: approximately 160 mm × 85 mm × 60 mm

The Leica 1 officially entered production at the end of 1924 and was the first stills camera successfully mass-produced for the 35mm cinema film format. Compact, fast, and mobile, it changed the role of the camera, democratising and modernising photography. It became a tool not just for studios, but for artists, journalists, and everyday people to document the world.

Oskar Barnack wanted a camera that 'didn't get in the way', letting photographers focus on the moment, not the workings of the machine. Its legacy continues to resonate, reminding designers and artists that careful consideration of form, function, and usability can transform the way society sees itself and the world.

Design & Features

The Leica 1 is highly portable, designed to be carried to the subject rather than bringing the subject into a studio. Its compact, metal-clad body is light enough to fit in a pocket, yet engineered to withstand everyday use in fast-moving environments. Controls are simple and direct: an aperture slider on the lens, a speed dial on the body, and an optical viewfinder for framing. Together, they allow the user to play with light, exposure, and composition without distraction. The lack of an integrated rangefinder in the Leica 1 meant focus was estimated, demanding immediacy over precision.

Its small size and low weight make it quick to handle, while 35 mm film cassettes let photographers carry multiple rolls with ease. This combination of portability, robustness, and straightforward controls turned photography into a more fluid and spontaneous practice.

Typical Workflow

1. Load 35mm film into the Leica 1's reusable cassette.
2. By eye, frame a subject in the optical viewfinder (V) by moving the camera or the subject.
3. Focus the lens (F) by estimating and setting the distance (range) between the camera and the subject.
4. Choose an aperture (A) and a shutter speed (S) to adjust depth of field, sharpness of movement, and exposure (using a separate light meter or learned rules of thumb).
5. Press the shutter to expose the film and take the photograph.
6. Manually advance the film, resetting the shutter, and be ready to capture the next picture.

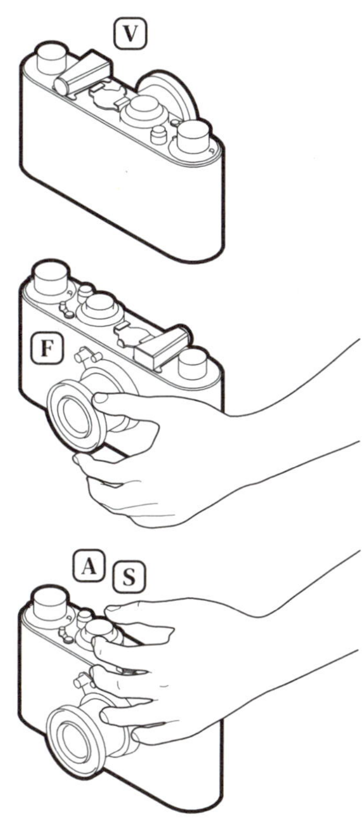

Collapsible Paint Tube Cultural Impact

The collapsible paint tube transformed artistic culture by altering where, how, and why painting took place. Making paint portable and immediately usable allowed artists to work directly in the real world rather than reconstructing it later from sketches and memory, placing speed, observation, and lived experiences at the centre of artistic practice.

The impact was most visible in the rise of French Impressionism. Working en plein air, Claude Monet, Pierre-Auguste Renoir, and Camille Pissarro all sought to capture transient light, weather, and movement as it unfolded. Rather than rendering objects in detail, they painted the impression of seeing, using rapid, broken brushstrokes to capture moments that vanished almost as quickly as they appeared. Monet's Impression, soleil levant (1872), painted at dawn in the harbour of Le Havre, depended on immediate colour decisions and swift brushwork to convey atmosphere rather than form. Renoir's Bal du moulin de la Galette (1876) extended this approach to modern urban life, recording moving crowds and shifting patches of sunlight. Renoir later acknowledged, 'Without paint in tubes there would have been no Monet, Cézanne, Sisley, or Pissarro.'[4] The screw-cap tube also enabled the plentiful use of vibrant pigments such as Chrome Yellow and Cobalt Blue. Post-Impressionist painter Vincent van Gogh exploited these in the thick, sculptural impasto of The Starry Night (1889), applying paint directly from the tube in a way that conveyed emotion as much as observation.

By democratising materials and removing dependence on the studio, the paint tube helped dismantle academic tradition, and a modern art emerged centred on immediacy, experimentation, and personal vision.

Singer Model 15 Cultural Impact

The Singer Model 15 reshaped domestic creativity, access to fashion, and the meaning of skilled labour in everyday life. Produced for over a century, it became a platform for sewists to develop techniques, confidence, and personal expression. It crossed generations and accompanied households through dramatically different social and aesthetic eras, from Edwardian formality to postwar reconstruction.

A standardised needle, bobbin, and tool-attachment system enabled home users to replicate professional techniques such as ruffling, pintucking, and fine topstitching. Garments with fashionable styles, once only produced in factories, could now be replicated at home, altering who could participate and how quickly new trends circulated. Performance with delicate fabrics such as rayon and silk proved especially influential in the 1920s. Home sewists could master the airy stitches and intricate surface detailing associated with the flapper dress and modern couture, moving away from heavy Victorian materials toward the fluid, draped silhouettes of the Jazz Age, helping modern fashions to reach a broader public.

The accessible interface of the Model 15 came to symbolise technological competence at a time when mechanical labour was increasingly valued, yet often excluded from women's work. It enabled many women to master intricate mechanical skills in the home, thereby helping shape modern perceptions of domestic labour and technical ability. Mary Brooks Picken, who founded the Woman's Institute of Domestic Arts and Sciences in Pennsylvania in 1916, taught millions of women using Singer standards and famously stated that the machine allowed a woman to 'Clothe her family with the same distinction as the wealthy.'[5]

Underwood No. 5 Cultural Impact

The Underwood No. 5 refined the mechanics of typing and, in doing so, transformed both the act itself and the culture of the early 20th century. Its visible front-stroke typing system encouraged a sense of forward momentum and narrative flow that went on to shape the structure and pace of modern prose. Often linked to Ernest Hemingway's concise, percussive style, the Underwood's rhythm demanded a firm, deliberate strike, producing a distinctive clatter that many writers came to adopt as a steady creative metronome.

Poets such as E. E. Cummings approached the Underwood as both a design tool and a writing machine. Its spacing, punctuation, and fixed-width typography enabled the page to be treated as a visual field, encouraging concrete and visual poetry and repositioning the typewriter as a medium for artistic experimentation. The machine also imposed a discipline that subtly shaped literary tone. Early models lacked a dedicated exclamation point, requiring typists to compose one by pressing a period, a backspace, and an apostrophe. The inconvenience discouraged emotional excess and instead rewarded precision. When combined with the difficulty of correcting errors, this produced writing that favoured clarity, restraint, and economical phrasing.

Its upright, open frame established the visual language of the modern office. Speed and reliability transformed clerical labour access to professional work, including suffrage organisations that relied on it to produce letters, circulars, and manifestos at scale. In wartime propaganda posters, the image of a female typist at an Underwood-style machine served as a symbol of national productivity and civic duty.

Leica 1 Cultural Impact

The Leica 1 expanded visual storytelling by making photography genuinely mobile. Its compact form encouraged photographers to step out of the studio and into the flow of everyday life, enabling candid images that felt immediate rather than staged, leading to a shift that supported the rise of photojournalism and street photography as significant artistic and cultural practices. Oscar Barnack himself demonstrated the camera's potential, documenting the 1920 flood in Wetzlar to reveal how a small, portable camera could capture fast-moving events with clarity and presence. Although Barnack was not a professional photographer, his work demonstrated how quickly the Leica could be used to record life as it unfolded.

Documenting human experience and natural events illustrated the camera's cultural power, inspiring photographers to explore everyday street scenes and social conditions. Henri Cartier-Bresson famously used a Leica to capture fleeting moments in Paris, creating compelling images of life moments that influenced generations of photographers. Robert Capa relied on Leicas to record war and social upheaval, helping to establish a documentary style grounded in proximity and emotional truth. Their images demonstrated how the camera's design encouraged engagement with the world rather than distance from it.

The camera also became a symbol of modernity and taste. In fashion and advertising, it enabled photographers to experiment with movement and candid expression, producing images that conveyed spontaneity and sophistication. Art photography practitioners such as André Kertész explored composition and perspective, observing scenes from locations and angles that had previously been difficult to achieve.

23 | RCA Theremin

Category: Music

Release: 1929 (1928 patent)

Designer: Lev Sergeyevich Termen (Leon Theremin)

Manufacturer: RCA (Radio Corporation of America), North America

Dimensions: approximately 485 mm × 1,187 mm × 308 mm (with stand)

The theremin was one of the earliest electronic musical instruments and remains one of the few played without physical contact. In 1920, Russian physicist Lev Sergeyevich Termen (Leon Theremin) was measuring gas densities when his hand affected the readings. Connecting an amplifier and speaker, he began playing musical phrases, and his discovery evolved into an instrument simple in form, elusive in mastery, and endlessly expressive. Its eerie, vocal-like sound came to define early avant-garde music and sci-fi cinema.

Though eclipsed for a time by more accessible electronic instruments, the theremin sound remains unique and has resurfaced in film scoring, experimental music, and popular culture. Replicas, modern kits, and open-source designs have broadened access and sustained interest.

Design & Features

The design is guided by the interaction between the performer's body and two antenna-based electromagnetic fields: a vertical rod governs pitch; a horizontal loop controls volume. As the performer moves their hands within each field, subtle shifts in capacitance alter the frequency of paired oscillators in an electronic circuit, converted into audio and amplified through an external loudspeaker. The instrument performs best when positioned in an open space, as each antenna responds to the surrounding environment and the performer, both of which become part of its functional design.

Slender legs raise the wooden cabinet, with facing knobs to trim pitch and volume, and toggle switches for muting and power. Despite its apparent simplicity, the instrument requires careful setup, warm-up time, and precise spatial awareness to achieve adequate control.

Typical Workflow

1. Step into position and let the instrument settle into a stable pitch field.
2. Move the volume hand to hover over the loop (L) and position it carefully to find the control boundary between quiet and tone.
3. Move the pitch hand near the vertical rod (R), listen and map out note intervals, using hand shapes and finger patterns to secure intonation across octaves. Make small hand oscillations to produce vibrato, and larger movements to glide between notes.
4. With the volume hand, try making 'pecking' gestures to articulate discrete notes. Part instrument, part performance, playing feels like conducting energy: the sound responds to motion, and the experience is intuitive because the mapping between gesture and result is direct and immediate.

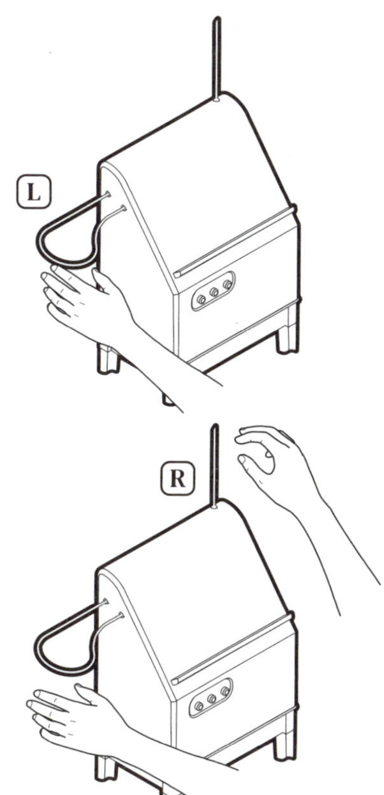

24 | Fender Stratocaster

Category: Music

Release: 1954

Designers: Leo Fender, George Fullerton, Freddie Tavares, Bill Carson

Manufacturer: Fender Musical Instruments, North America

Dimensions: approximately 324 mm × 984 mm × 44 mm

Introduced in 1954, the Fender Stratocaster has become one of the most influential design objects in modern music. Its sculpted double-cutaway form, trio of single-coil pickups, and remarkably versatile tonal palette set a new standard for electric guitar design.

Leo Fender, a radio repairman rather than a guitarist, had an engineering background, giving him a distinctive point of view that helped him to imagine an instrument shaped by technical ingenuity as much as artistic ambition. Instead of relying on craft tradition, he approached specific needs voiced by musicians. The result addressed long-standing issues of comfort, balance, playability, expressiveness, personalisation, and ease of maintenance, creating a guitar purpose-built for the evolving demands of modern players.

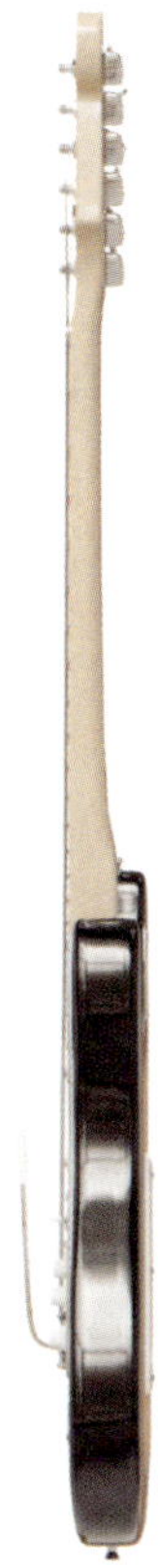

Design & Features

A contoured back and carefully balanced weight distribution provide playing comfort, making the Stratocaster a preferred choice for extended live performance. The double cutaway improves access to the upper frets, expanding the melodic range, and the volume and tone controls are placed so they can be adjusted easily with an extended finger during play. Three single-coil pickups can be combined in various configurations through a five-way selector switch, offering broad tonal flexibility.

The tremolo arm system introduces a pivoting mechanism that operates on a fulcrum, a significant change from earlier fixed bridges. By varying string tension, it turns pitch into a dynamic, expressive performance variable. A bolt-on neck, accessible electronics, and straightforward internal layout make an inviting platform for modification and adaptation.

Typical Workflow

1. Find a voice that suits the musical idea you want to develop. Try the five-way pickup selector (P): the bridge offers a bright character suited to rock and country, the neck a warmer quality associated with blues or jazz, and the intermediate position delivers a slightly hollow tone suited to funk and clean pop.
2. Let the instrument's physical response shape your playing. Use expressive techniques such as vibrato.
3. Once a promising riff, chord sequence, or melodic line appears, refine it. Subtle adjustments to the tone control (T) can give each section a distinct identity. The Stratocaster's dynamic range ensures that small changes in picking intensity or control settings produce significant differences.

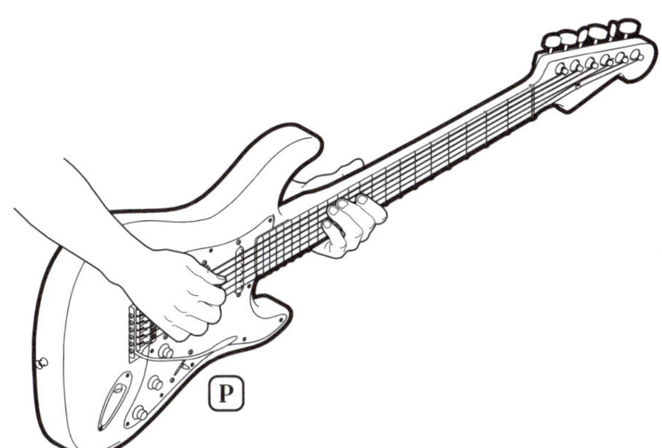

RCA Theremin Cultural Impact

Although RCA marketed the theremin as 'A musical instrument anyone can play', it required an exceptional ear and extensive practice, offering no tactile or visual cues for locating notes. Yet its brief commercial life sparked a new era in electronic sound. Defying conventional performance paradigms, musicians shaped pitch and volume without touching the instrument, achieving a vocal-like expressiveness and vibrato that Clara Rockmore described as 'Delicate motions, you have to play with butterfly wings.'[6] Rockmore, a classically trained violinist, elevated the theremin from novelty to a serious instrument through her performances and recordings, including The Art of the Theremin (1977), demonstrating a high level of technical and musical sophistication.

In Hollywood, the professional player Dr Samuel J. Hoffman introduced the theremin to film composers, with his ethereal performances for Alfred Hitchcock's Spellbound (1945) and The Day the Earth Stood Still (1951) becoming a shorthand for psychological tension and science-fiction atmospheres. Composer Bernard Herrmann's use of two theremins in The Day the Earth Stood Still heightened the film's eerie, otherworldly character, associating the theremin sound with the supernatural and the futuristic.

Influence also reached popular and rock music. The Beach Boys' Good Vibrations (1967) employed an electro-theremin (with a ribbon controller and a slider, played more like a trombone), and Jimmy Page of Led Zeppelin incorporated one into live performances, showing its ongoing role in sonic experiments. Robert Moog, who built theremin kits as a teenager, credited the instrument as foundational to his innovations in analog synthesisers and the modern electronic sound.

Fender Stratocaster Cultural Impact

The Fender Stratocaster moved the electric guitar from a background rhythm instrument to a foreground tool for personal expression, identity, and performance. When Buddy Holly appeared on American television in the 1950s, its space-age contours and bright visual language marked a clear break from tradition, establishing rock and roll as an optimistic, high-energy cultural force.

That symbolic power expanded dramatically in the 1960s. Jimi Hendrix played a right-handed Stratocaster upside down, altering its string tension and pickup response to pioneer new psychedelic sounds. Hendrix described it as an extension of his own body. His radical use of the tremolo system introduced siren-like sounds that oscillated between chaos and beauty, transforming the guitar into a tool for political protest and cultural commentary.

In the 1970s, the Stratocaster's practicality and versatility supported a range of musical movements. Nile Rodgers exploited its sharp, percussive clarity in the rhythmic foundation of disco, while Eric Clapton praised it for blues-driven rock, saying, 'My feelings about a perfect design is that it has to be functional. That's what makes the design so beautiful. It's superbly thought out.'[7] Modular bolt-in components allowed Eddie Van Halen to build his 'Frankenstrat', which fused Stratocaster ergonomics with high-output humbucking pickups, catalysing the aggressive tone that defined 1980s hard rock and metal.

Through these musicians and countless others, the Stratocaster has shaped the sound, aesthetics, and working methods of rock, blues, funk, and beyond. It is a global design archetype that continues to influence how personal creativity is performed, embodied, and heard.

25 | Kinematograph

ca. 1890 — The first motion-picture cameras used a sprocket system and perforated celluloid film to record and project moving images.

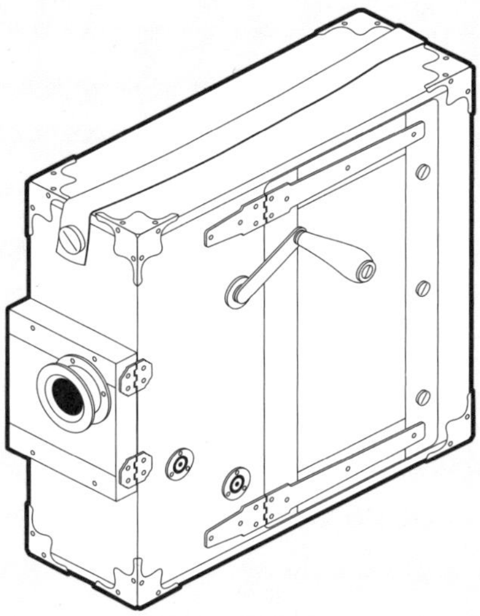

26 | Bauhaus Course

1919 — The Bauhaus Preliminary Course broke from academic tradition. Instead of mastering predetermined styles, first-year students explored colour, form, material, texture, and perception through hands-on experimentation. Its roots lay in a fusion of craft education, modern art, and industrial thinking, influenced by movements such as Expressionism and Constructivism.

It reminds us that tools do more than enable output; they shape how we think. By foregrounding process, play, and sensory understanding, the Bauhaus set a template for creative systems that empower exploration rather than prescribe outcomes, thereby making the idea as relevant today to designing physical and digital tools as it was in studios and workshops a century ago.

27 | X-Acto Knife

1930 — Defined by a screw-clamp handle and replaceable blades, a design that has remained largely unchanged for nearly a century.

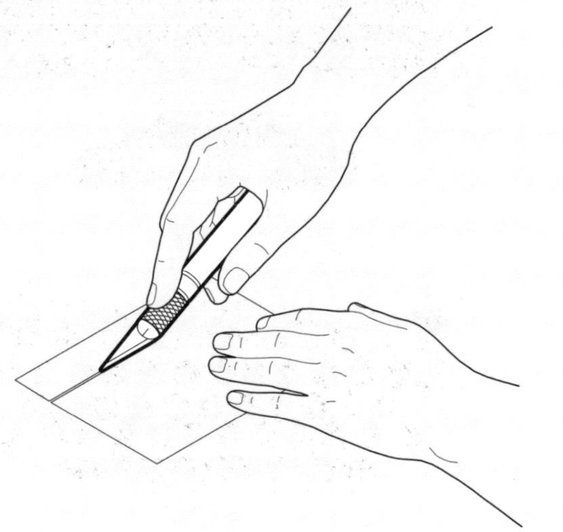

28 | Lettera 22

1950 — The Olivetti Lettera 22 looked like the future on arrival. Designed by Italian industrial designer and architect Marcello Nizzoli, it rejected the heavy aesthetic and exposed mechanics of earlier typewriters in favour of soft edges, a low profile, and modern colours: a machine that felt approachable rather than intimidating.

Nizzoli was central to Olivetti's shift toward modernist principles, introducing clarity, proportion, and restraint at a time when office tools were ornate and authoritarian. For designers today, the Lettera 22 shows how visual calm and friendliness can extend a tool's reach and use.

◂ *Olivetti Lettera 22 Typewriter (1950)*

29 | CNC Machine

1952 — Computer Numerical Control (CNC) was introduced, with motor-driven machine tools guided by punched tape programs, moving beyond the limits of mechanical, cam-based automation. Drawing inspiration from the Jacquard loom (ca. 1804), Thomas Blanchard's gun-copying lathe (1818), and Charles Babbage's mechanical computing (ca. 1820), CNC brought abstract programmability into industrial fabrication.

Designers could now translate complex geometries into precise, repeatable machine motions. Microprocessors in the 1970s and subsequent reductions in computing costs have enabled multi-axis milling, sheet laser cutting, and low-cost 3D printing.

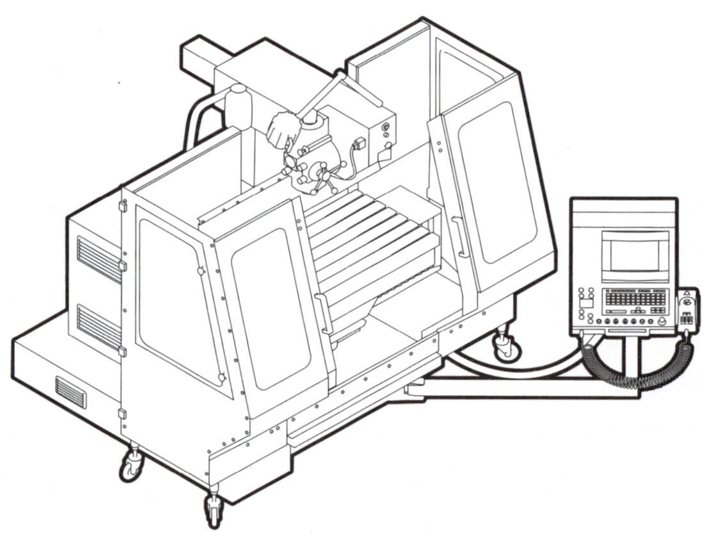

30 | Moog Synthesiser

1964 — Robert Moog, a physicist and electronics enthusiast known for building theremins, met composer Herbert Deutsch, who encouraged him to design electronic circuits that musicians could control expressively, leading to the creation of the voltage-controlled Moog modular synthesiser. This radically new instrument was built from individual sound modules (comprising oscillators, filters, envelopes, and amplifiers) that could be connected and rearranged with signal patch cables. Instead of offering fixed sounds, the Moog allowed musicians to assemble their own configurations, making the instrument a reconfigurable, expandable, and open-ended system of sound construction.

▲ *Minimoog Voyager Synthesiser used by J Dilla (2002-2005)*

American classical composer and electronic music pioneer Wendy Carlos[8] worked closely with Moog, providing practical feedback that shaped key design features. Most notably, she influenced the inclusion and refinement of the piano-style keyboard, which made the synthesiser far more accessible to traditional musicians.

The Moog's legacy continues to influence contemporary practice, demonstrating how flexible systems and tactile interfaces can shape the success of creative tools. In the 1990s and early 2000s, American hip-hop producer and composer J Dilla brought the Moog into a new era, using it to craft thick, rounded basslines and subtly unstable tones that complemented his off-kilter rhythms. His use of the synthesiser highlighted its versatility beyond classical or experimental music.

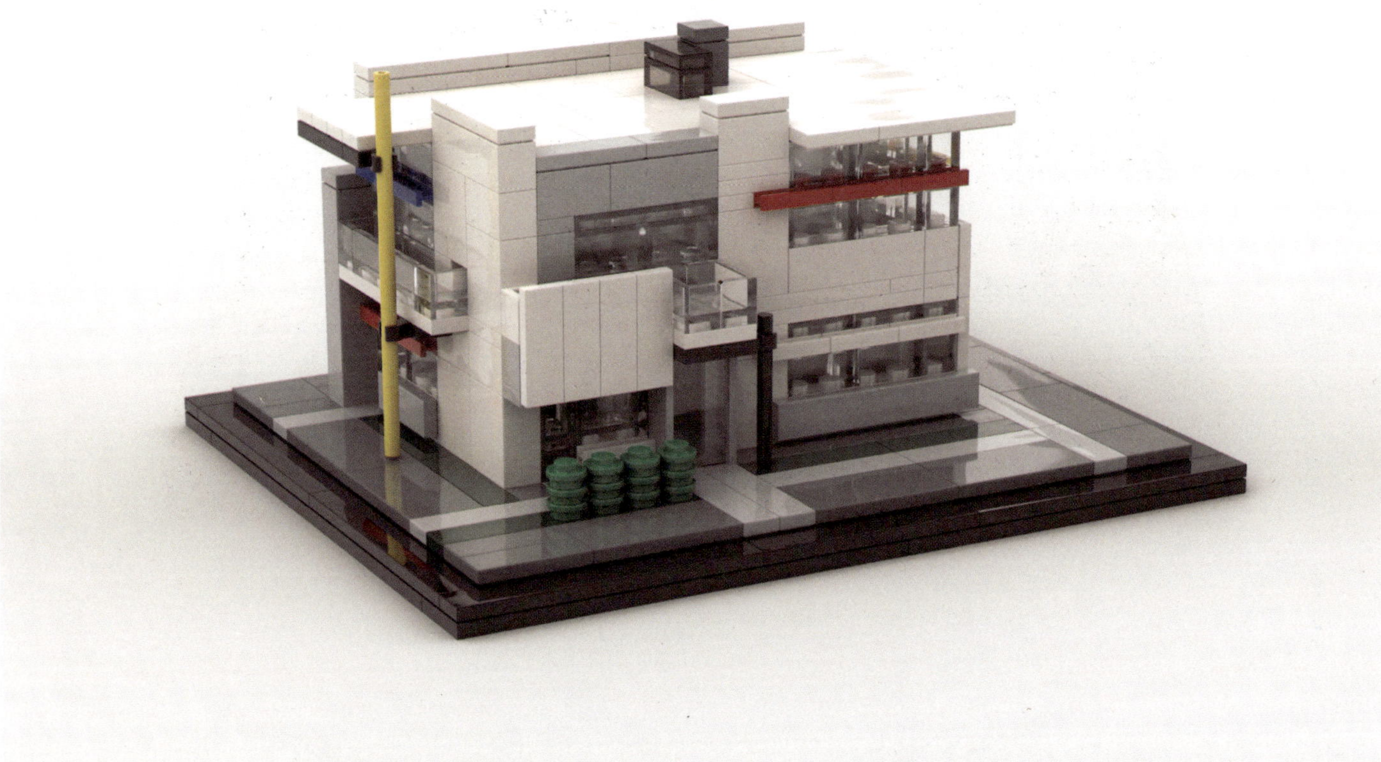

31 | LEGO Brick

Category: Play, Design

Release: 1958, 1977 (Technic brick)

Designer: Godtfred Kirk Christiansen

Manufacturer: LEGO Group, Denmark

Dimensions: approximately
32 mm × 11 mm × 16 mm
(2x4 brick)

Since the first plastic brick was patented in 1958, LEGO has become a universal design system used by children for imaginative play and also by adults, educators, architects, and designers as a medium for creativity, problem-solving, and storytelling.

LEGO empowers people of all ages to think with their hands, to turn imagination into something tangible, and to explore how structure, colour, and form interact. Over decades, it has transcended toy status to become a symbol of creativity, collaboration, and the enduring human desire to build, experiment, and play.

There are over one trillion LEGO bricks in the world. That's over 100 each for every human currently on planet Earth!

Design & Features

The LEGO brick is a masterclass in modular design. A simple, precision-engineered building system that balances constraint and possibility. Each brick connects with every other piece made since 1958, thanks to its exact design tolerances, material properties, and the stud-and-tube coupling system. This commitment to long-term interoperability creates a design ecosystem that evolves without obsolescence, a rare and powerful concept.

A satisfying physical 'click' provides feedback that connects the mind and the hand. A tactile loop of experience, flow, and play deepens engagement and enables the efficient building, modification, deconstruction, and rebuilding of a vast and diverse range of models and creations.

Builds can follow instructions, or ideas can be sketched and explored as digital 'mockups' with interactive software like LEGO Digital Designer.

Typical Workflow

A guided build comes with instructions:

1. Unbox and sort the bricks by colour and type (or you might choose to skip sorting).
2. Follow the build instructions sequentially. LEGO books are visual, showing where each brick attaches, step by step.
3. Watch the build take place, and then admire and play with the completed model.
4. Display, modify, and remix with other sets, or disassemble for future builds.

A creative build uses free and leftover bricks:

1. Decide what to build, or start without a plan and see how creativity emerges.
2. Try arrangements, colour schemes, and mechanics to tell your own stories.

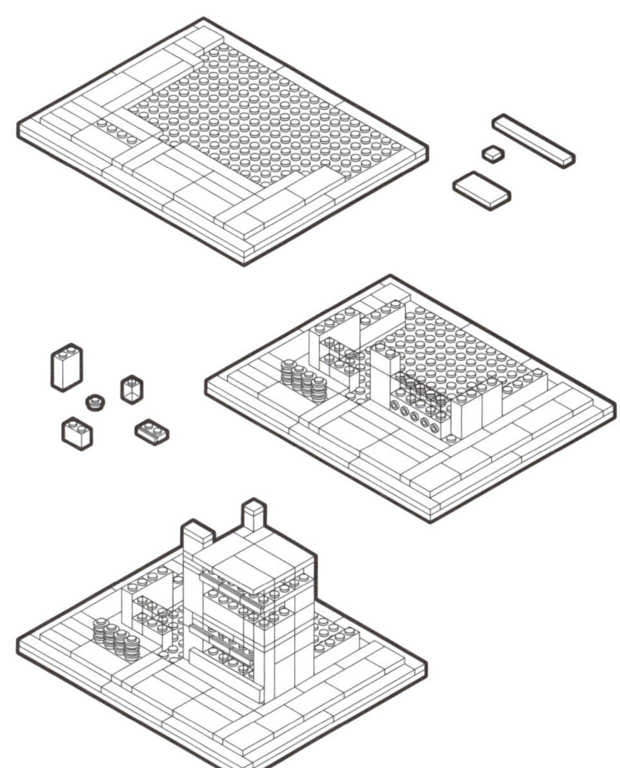

32 | Technics SL-1200

Category: Music

Release: 1972

Designer: Shuichi Obata

Manufacturer: Matsushita Electric, Japan (now Technics by Panasonic)

Dimensions: approximately 453 mm x 353 mm x 169 mm

Launched as a consumer hi-fi component, the SL-1200 turntable unexpectedly became the foundation of global DJ culture. Its reputation was built on a pioneering direct-drive system that delivered precision and torque that belt-driven decks could not match.

Unwavering performance made it essential to the emergence of hip-hop, disco, house, and techno, with DJs such as Grandmaster Flash and Frankie Knuckles transforming the SL-1200 from a playback machine into a creative tool, developing scratching, beat-matching, and looping as new forms of musical expression. Its layout and aesthetic continue to be referenced and reinterpreted throughout the industry, with many DJs still citing the SL-1200 as foundational to their craft.

Design & Features

A quartz-locked, direct-drive motor delivers high torque to the platter, eliminating belt stretch and slippage. The result is a near-instant start-up and the rotational accuracy required for precise cueing and mixing. A heavy, die-cast aluminium body and shock-absorbing base minimise resonance and vibration, giving the turntable impressive stability, even in the most energetic environments.

The pitch control slider, originally intended for technical calibration, allows DJs to synchronise the tempos of different songs, enabling seamless transitions and rhythmic manipulations. Later iterations add a reset button that immediately returns to the exact pitch. The SL-1200 family continues to refine both performance and ergonomics while preserving the familiar geometry that defines its interface and identity.

Typical Workflow

1. Line up a pair of SL-1200s and make your vinyl record choices.
2. Cue a track on each and monitor the incoming songs through your headphones.
3. Fine-tune the pitch slider (P) to align tempos, using its precise control.
4. With the beats locked in, use a wrist flick, a deliberate needle drop, or a quick platter stop to create rhythmic accents. Scratch, back-spin, or drop beats, feeling the tactile response and the subtle resistance of the platter.
5. As the room reacts, stay alert to the energy around you! Shape the mix in real time, moving intuitively from one transition to the next, sculpting the experience through sound, touch, and rhythm.

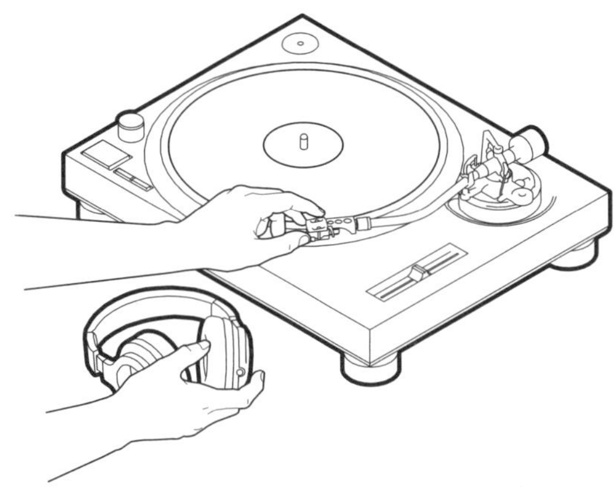

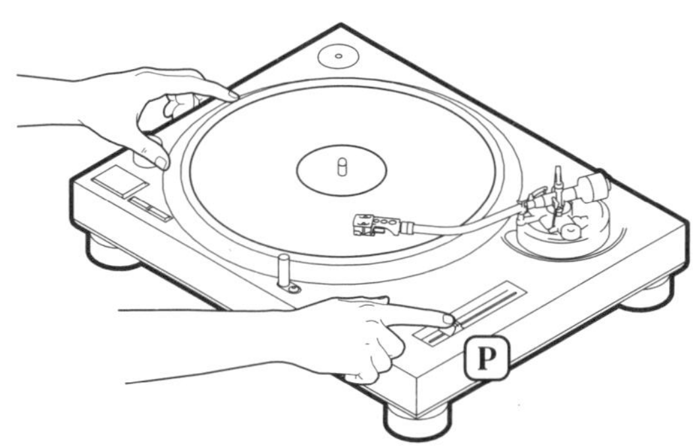

33 | Korg MS-20

Category: Music

Release: 1978

Designer: Hiroaki Nishijima

Manufacturer: Korg, Japan

Dimensions: approximately
573 mm × 248 mm × 316 mm

When released in 1978, the Korg MS-20 was one of the first accessible semi-modular synthesisers. It would produce sound immediately, but the patch panel of cables allowed musicians to override the internal wiring and reconfigure the signal flow themselves. This offered some of the freedom of a fully modular system without the cost or bulk, putting sound design directly in reach as a hands-on, visible process.

Compact, affordable, and distinctive in tone, it opened electronic music to a broader audience. With sharp filters and tactile design, it became a signature for artists such as A Flock of Seagulls, Orchestral Manoeuvres in the Dark, Portishead, Aphex Twin, and Daft Punk, securing its place as one of the most influential synthesisers of its era and among artists that followed.

Design & Features

An angled front panel places all controls and the patch panel within direct reach, making synthesis feel immediate and physical, and close to the piano keyboard.

The interface is a functional part of the machine. Each knob generates and shapes the sound output by directly controlling voltage, current, and resistance in electrical circuits that drive oscillators, filters, and envelope shapes. The patch panel encourages experimentation by letting musicians reroute signals in unconventional ways. Plugging and unplugging cables gives a direct, visual map of the sound architecture and signal flows.

Despite its compact size, the instrument is rugged enough for stage use. Decades later, an official reissue by Korg and the creation of build-it-yourself kits highlight just how enduring and adaptable the design has remained.

Typical Workflow

1. Select a basic oscillator waveform (W) to establish the sound's foundation.
2. Shape the character using the filters (F), adding bite or smoothness with cutoff and resonance controls.
3. Adjust envelopes (E) to control how the sound attacks, sustains, and fades. Controls are analog with no discrete steps or resolution limits. Sweeping smoothly through values can contribute to characteristic organic sounds.
4. Experiment with patching cables (P), connecting and rerouting signals to add modulation and feedback effects. Subtle, unexpected interactions between circuits can yield unique and interesting results.

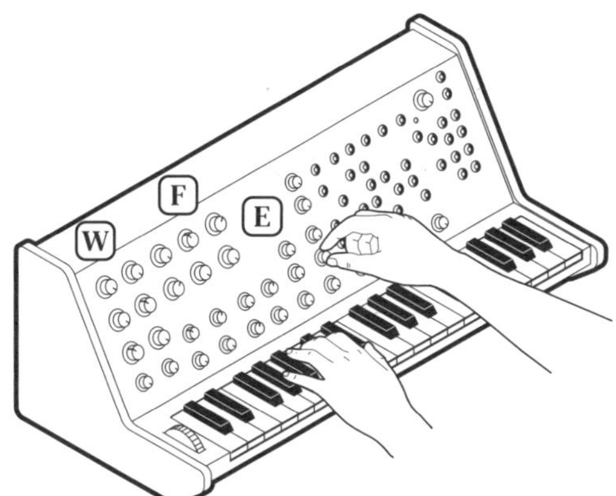

34 | Roland TR-808

Category: Music

Release: 1980

Designers: Ikutaro Kakehashi, Tadao Kikumoto

Manufacturer: Roland Corporation, Japan

Dimensions: approximately 508 mm × 305 mm × 105 mm

At the time, an affordable alternative to high-end drum machines, the TR-808 created sounds through analog synthesis rather than recorded samples. It was among the first instruments capable of programming and playing back complex patterns without a live drummer, yet its initial reception was muted. Artists and producers who prioritised realism dismissed its synthetic sound, and it was discontinued in 1983.

Over time, that same idiosyncrasy revealed a distinctive character: warm, sizzling, and slightly unstable, anchored by its iconic booming bass drum and crisp hi-hat. The 808 went on to feature in more hit records than any other drum machine, its influence stretching from early hip-hop and electro into house, techno, and pop. It remains widely used in original, official reissue, and software forms today.

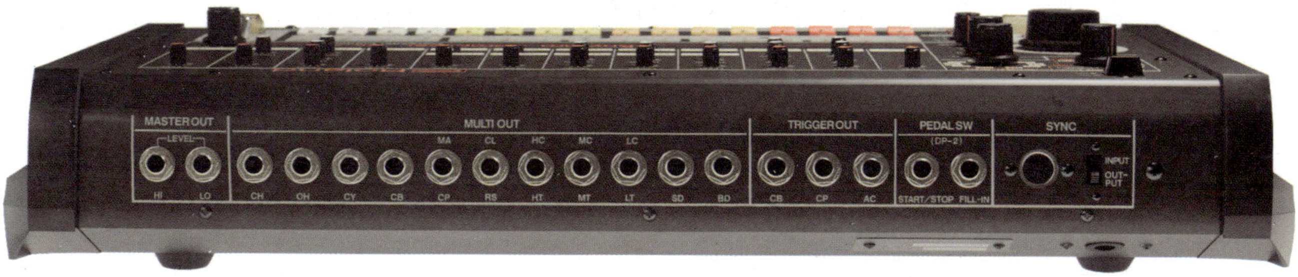

Design & Features

Synthetic drum sounds are generated through a combination of oscillators, noise sources, filters, and envelope-shaping electronic circuits. Because these are analog and rely on imperfect and individually unique out-of-specification transistors, each sound exhibits subtle variations shaped by age and room temperature, along with built-in controls for tuning, tone, and decay.

The interface for playing and programming is immediate and visually clear. Hit buttons align with grouped parameter knobs, and pattern-based light feedback creates an intuitive workflow that encourages live loop-based improvisation. A design language of beat steps, per-voice controls, and pattern chaining became the template for many later electronic music tools. Early sync and trigger-out ports (appearing before MIDI standardisation) enable the 808 to coordinate and trigger other devices, extending its role in the studio and on stage in live performances.

Typical Workflow

1. Select one of 32 pattern slots and a drum voice (V): bass drum and snare, toms, conga, rimshot, claves, handclap, maraca, cowbell, cymbal, and open or closed hi-hat.
2. Hit the 16 individual step buttons (B) to place beats, using the grid layout and light feedback to visualise timing. Hit START (S) to loop the pattern and hear the selected steps play back.
3. Toggle steps while the sequence runs, programming the rhythm in real time. Layer additional drums and adjust the levels, tones, tunings, and decays (A), and the overall tempo (T).
4. Chain completed patterns into extended song sequences. Get creative and treat the machine as an instrument in its own right rather than simply a drummer stand-in.

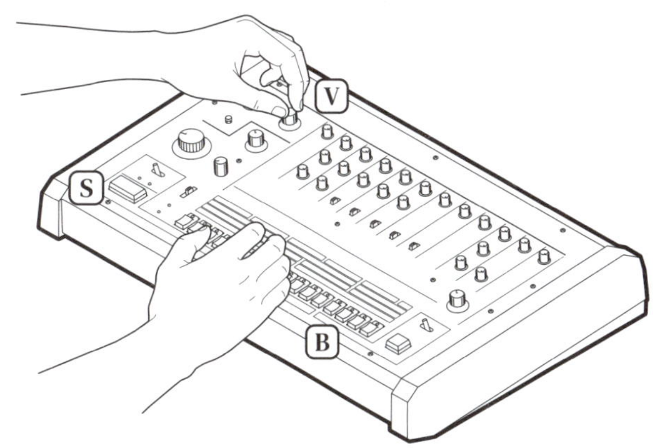

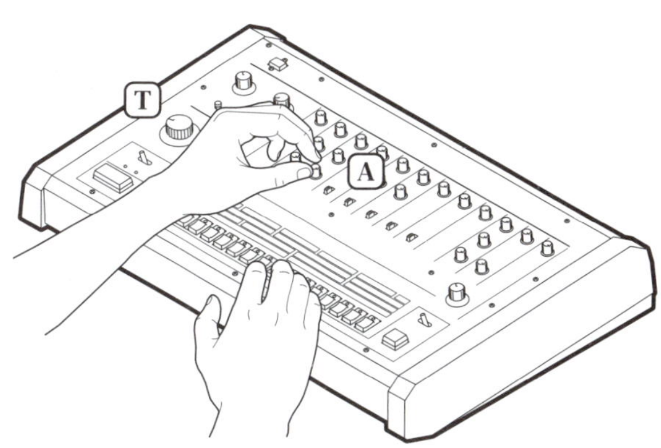

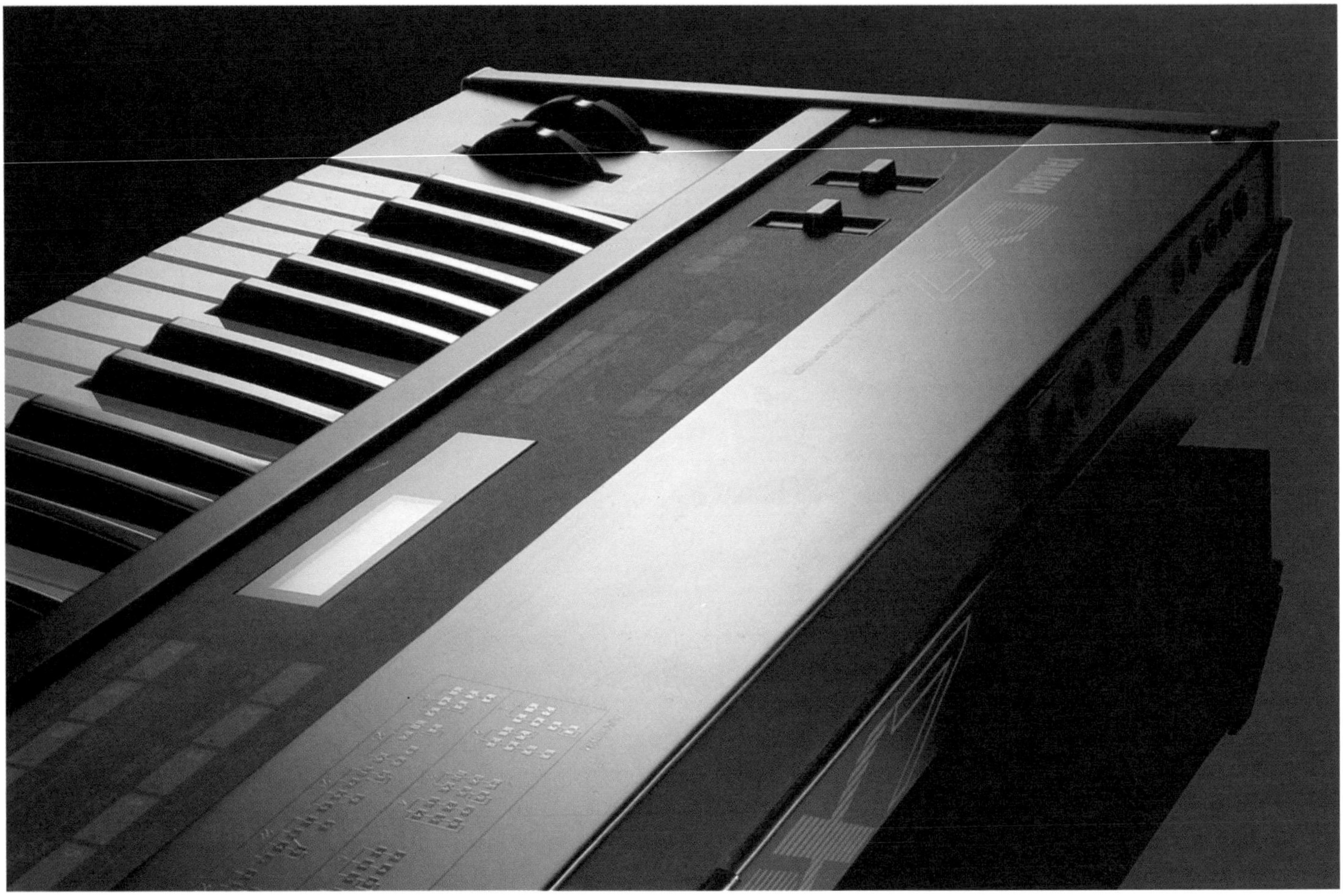

35 | Yamaha DX7

Category: Music

Release: 1983

Designers: Ed Cornett, John Chowning, Dave Bristow, Gary Leuenberger

Manufacturer: Yamaha, Japan

Dimensions: approximately 1,018 mm × 102 mm × 239 mm

An unassuming keyboard, the Yamaha DX7 was the first truly mass-market digital synthesiser. It reshaped the sound of popular music in the 1980s with crisp electric pianos, glassy bells, and metallic basses, heard on everything from chart-topping pop culture and R&B to film soundtracks.

It implemented Frequency Modulation (FM) synthesis, a new method for generating sound discovered in the 1960s by Stanford researcher John Chowning. Unlike analog synthesisers, which began with harmonically rich waveforms and then filtered them down, FM synthesis built complexity by modulating simple sine waves against each other. The result was a palette of bright, dynamic tones with a precision and clarity no analog system could match.

Design & Features

Yamaha licensed FM synthesis and turned academic theory into a practical, affordable instrument, not attempting to mimic nature exactly, but creating a new sonic language. Choosing to embrace digital precision marks a turning point when music shifts from tactile and handmade to calculated and programmable. Custom silicon chips make it affordable and portable for the first time.

The visual style and interface language suggest order, efficiency, and technological confidence. Sound design is handled through a small LCD with layered menus controlled via touch parameter selection and a data-value input slider. New sounds are notoriously tricky to program, so few musicians venture beyond the widely heard, now-famous factory presets.

The DX7 was among the first to support built-in MIDI (Musical Instrument Digital Interface), a standard that enables instruments and computers to communicate note and sound data. It represented a philosophical shift toward interconnectivity among tools, laying the groundwork for the networked world of modern music technology.

Typical Workflow

1. Select a factory preset voice (V): Piano, Marimba, Tubular Bells, or Synth Lead!
2. Play the velocity- and touch-sensitive keys, using variations in speed and pressure to affect brightness, volume, and attack.
3. Feeling more adventurous? Enter sound edit mode using the LCD screen (S), touch buttons, and the data input slider (D). Each sound is built from six operators: simple sine waves, combined in 32 possible ways, referenced in small diagrams reproduced on the front panel.

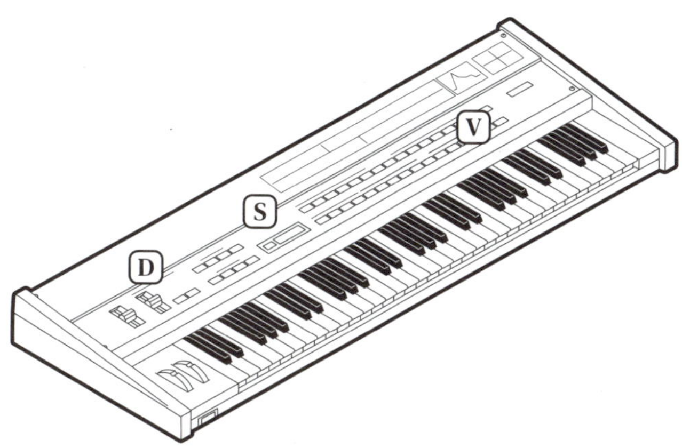

36 | Wacom Series

Category: Photography, Illustration, Design, Animation

Release: 1984, 1998 (Intuos series), 2010 (Cintiq series with built-in display)

Designer: Masahiko Yamada

Manufacturer: Wacom, Japan

Dimensions: approximately 291 mm × 7 mm × 206 mm (Intuos Pro Medium)

Wacom set out to make digital creation feel as intuitive as drawing on paper. At a time when computers were primarily controlled with keyboards and mice, the Wacom WT-460M became the world's first tablet to feature a cordless stylus, a breakthrough that opened the door to a new kind of creative interaction between hand, eye, and screen.

From its early adoption by Walt Disney Studios for the animated feature Beauty and the Beast in 1990 to its widespread use across design, photography, film, and animation production, Wacom evolved from a niche tool into an essential interface for digital art. Its appeal lies in the subtlety of control and the familiarity of movement, pressure, and gesture, which successfully bridge traditional artistry and digital creation.

Typical Workflow

1. Set up and calibrate the tablet by adjusting pen pressure and tilt response, and mapping touch and key shortcuts to the most used software actions.
2. Hold the pen naturally and use it to draw and design on the tablet. Over time, muscle memory will develop, allowing your hand to move freely and accurately across the surface while you keep your focus on the screen.
3. Use pressure and tilt to create exploratory and expressive marks with controlled variations in line, form, thickness, and texture.
4. Use the on-pen and tablet context buttons (B) to switch instantly between brush styles and eraser types.

Design & Features

Wacom's design philosophy centres on maintaining precise control and creative flow without interruption. Electromagnetic resonance technology allows the pen to operate without batteries or cables, drawing power wirelessly from the tablet's magnetic field. This makes it lighter, more precise, and highly reliable. The pen senses pressure, tilt, rotation, and velocity, capturing the subtle dynamics of drawn strokes.

Programmable Touch Strips and Express Keys put essential controls such as zoom and brush size directly at the artist's fingertips, without needing to look away from the screen or reach for a keyboard.

Wacom's hardware shapes the evolution of creative software itself. Software applications like Photoshop adapt to interpret pressure and tilt, allowing digital brushstrokes to behave in a responsive, organic, and artistic way.

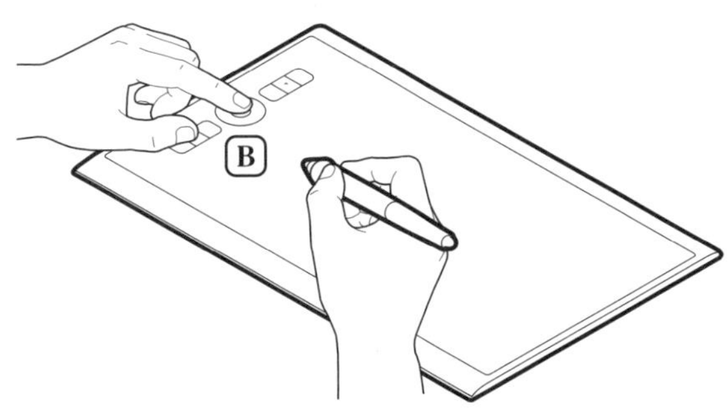

LEGO Brick Cultural Impact

The LEGO brick stands as one of the most well-known and widely used design systems of the 20th century. Its stud-and-tube connection creates a precise balance between rigid standardisation and virtually infinite possibility, rewarding problem-solving and imagination over passive consumption.

The introduction of the minifigure in 1978, designed with a neutral expression and a human scale, encouraged storytelling without prescribing it. LEGO's early lack of narrative specificity made the brick inherently gender- and age-neutral, allowing anyone to project themselves into the system. Compatibility across generations reinforced this effect, turning LEGO into a lifelong tool rather than merely a childhood phase. A pivotal development occurred in 1999 with the introduction of the licensed Star Wars franchise, followed by Harry Potter in 2001. By translating well-known story worlds into modular form, LEGO invited fans to reconstruct their own narratives and remix universes in a single build.

Artist Nathan Sawaya demonstrated how a simple plastic system could convey emotion through large-scale sculptural works displayed in public exhibitions such as The Art of the Brick. He observed that the brick's accessibility enabled viewers to connect strongly with his art because they have a personal history with the material.

LEGO Serious Play brought bricks into corporate boardrooms as a tool for building metaphors and flattening hierarchies. LEGO's impact lies in a simple system that reshaped play, storytelling, learning, collaboration, and design thinking itself. Never merely a toy, it is a universal medium for expression in which a fixed system becomes the ultimate catalyst for freedom.

Technics SL-1200 Cultural Impact

Emerging in the Bronx, New York, during the early 1970s, hip-hop pioneers such as DJ Kool Herc, Grandmaster Flash, Grand Wizzard Theodore, and Afrika Bambaataa relied on the SL-1200 at block parties in parks and on closed city streets, where community gatherings turned spaces of urban decline into sites of cultural experimentation. Its durable design and stable, direct-drive motor make it well-suited for carrying and use outdoors.

The Bronx cultivated hip-hop's core practices of DJing, MCing, breakdancing, and graffiti, which together offered creative alternatives to gang violence and fostered unity, identity, and self-expression among marginalised youth. Kool Herc developed breakbeat looping; Grandmaster Flash advanced technical precision, using slipmats to reduce friction and quick-mix theory to transition between records; Grand Wizzard Theodore introduced scratching; and Afrika Bambaataa mobilised hip-hop as a force for community cohesion. Central to this experimentation was the turntable itself.

During the 1980s and 1990s, the SL-1200 became a global standard in clubs. Its cultural reach expanded through the DMC World DJ Championships, which established *turntablism* (the performance art of manipulating sounds and creating new music with turntables). Through beat juggling, tone play, and increasingly complex scratching techniques, competitors demonstrated that two SL-1200s and a mixer could rival the musicality and impact of any traditional musical instrument. Today, the SL-1200 stands as a symbol of innovation shaped by performance and resilience, and as a creative tool that has helped carry a local identity and street culture onto the global stage.

Korg MS-20 Cultural Impact

Establishing itself in the 1970s as a defining sound engine in electronic and pop music through its raw analog tone, semi-modular patching system, and uncompromising filters, the MS-20 rewarded bold sonic and musical experimentation. By the early 1980s, it had become integral to an emerging synth-pop style. A Flock of Seagulls employed it to drive the bass line and futuristic textures in 'I Ran', contributing to a distinctive new wave sound. Orchestral Manoeuvres in the Dark incorporated the MS-20 on songs like 'Bunker Soldiers', underpinning synth-pop's melodic yet mechanical and analog aesthetic.

Outside of its own sound, the inclusion of an external signal processor enabled artists to treat it as a transformative tool for other sources, leading to unexpected creative shifts. French duo Air used the device to process vocals on their album Moon Safari[9]; Aphex Twin integrated it into experimental and ambient techno contexts, often routing external sounds to blur the lines between synthesis and acoustic manipulation; and Portishead's guitarist and producer Adrian Utley processed guitar sounds, adding haunting colour to the band's pioneering trip-hop style. In the 1990s and beyond, the gritty, analog presence resurfaced within French house and techno production. While gear historians debate the exact instruments used by Daft Punk, the MS-20 influence is clearly evident in aggressive bass textures and tonal contours of early tracks such as 'Da Funk'.

Across each era, the MS-20's design limitations proved to be a creative asset. Its monophonic architecture and lack of memory pushed players to perform live, exploring instability, noise, and feedback as musical elements that influenced the production and performance of raw electronic music.

Roland TR-808 Cultural Impact

Initially criticised for its unrealistic-sounding drums, the TR-808 nevertheless attracted early adopters who explored its creative possibilities. When the experimental Japanese electronic group Yellow Magic Orchestra brought it onstage in 1980, they introduced audiences to a futuristic, synthetic vibe not heard before. Afrika Bambaataa and Arthur Baker used the TR-808 to create the 1982 hit 'Planet Rock'[10], a song that defined electro-funk and demonstrated significant potential for musical innovation. Marvin Gaye later showcased its adaptability on 'Sexual Healing', using the machine's crisp handclaps and hi-hats to generate a sensual, steady pulse, proving that a departure from realistic sounds could be a unique advantage.

Hip-hop artists Run-DMC, LL Cool J, and Public Enemy, and Def Jam Records producer Rick Rubin, all relied on the TR-808's unique low-frequency kick drum to drive beats. Public Enemy producer Hank Shocklee emphasised that the machine's synthetic quality was precisely what made it superior, sampling its sounds from existing recordings to achieve a specific, dirty quality that fit their aesthetic.

Detroit pioneers, including Juan Atkins, used its programmable patterns to build the tight rhythmic frameworks of techno, while Chicago house producers emphasised its sounds to shape dance music. The deep bass response and prominent snare were defining elements of Miami bass culture, and later, influenced trap music, in which the kick drum was tuned to function almost like a melodic instrument.

Today, the term '808' has shifted from a model number to a widely used descriptor for the deep, floor-shaking bass that drives mainstream EDM in dance clubs and at festivals worldwide.

▲ *Daft Punk after the release of Random Access Memories*

Yamaha DX7 Cultural Impact

The Yamaha DX7 offered something markedly different, standing out from the warm analog sound that had dominated synthesisers through the 1970s. Shorthand for the 'Sound of the 80s', it was so distinctive that later decades viewed its precise, metallic basses and dynamic electric piano presets as emblematic of the era. In 1986, the electric piano sound reportedly appeared in around 40% of US Billboard Hot 100 number-one songs, with artists such as Tina Turner, Michael Jackson, Whitney Houston, Phil Collins, Billy Ocean, Celine Dion, and George Michael all using the DX7 to create a clean pop and ballad sound, defining the decade.

Stevie Wonder used the DX7, mastering its expressive, velocity-sensitive response in funk and R&B songs. The way it responded to touch helped push electronic keyboard performance toward a more dynamic, player-centric approach. Jan Hammer used one for expressive lead sounds in synth-pop and TV soundtracks, exploiting the pitch and modulation capabilities to mimic guitar-like phrasing. Both musicians demonstrated that synthesisers were now a viable and interesting alternative to traditional instruments in lead parts.

The minimalist interface, small display, and mathematical logic made programming new sounds particularly difficult, leading to the overuse of factory presets and a homogenisation of the pop-music sound. Brian Eno took time to master its limitations, driving the complex menus as an engine for his ambient soundscapes. This required a different type of discipline, shifting from physical manipulation toward a more conceptual, calculated approach that carried over into the digital, menu-driven, computer-based workflows that propelled digital music production into the 1990s.

Wacom Series Cultural Impact

Wacom became essential during the transition from pencil sketching and inking to computer-based production. The adoption of the pen tablet by major film and animation studios transitioned digtial creative workflows from mouse-based input to a more natural interaction, closer to traditional hand-drawing. By the early 1990s, hand-drawn lines integrated directly into digital systems, preserving quality, enabling faster colouring workflows and allowing for more complex animated and simulated visual effects.

In television production and motion graphics, pen tablets became a valuable tool because artists could work intuitively with an absolute one-to-one mapping from hand input to images and motion paths. Graphic designers using software such as Creative Suite found that they could control brushes, curves, and shapes more intuitively and fluidly than clicking and dragging with a mouse. In industrial design and CAD environments, pen tablets integrated with tools such as SOLIDWORKS enabling designers to draw splines, handwrite dimensions, and manipulate 3D surfaces directly, allowing engineers to feel more connected to the flow of their sketches and the complex surfaces of their products.

The ergonomics of using a pen as the input device reduced repetitive strain and wrist fatigue compared with prolonged mouse use, and absolute input mapping allowed the cursor to jump directly to precise screen coordinates (whereas a mouse must be moved relative to its last position). Experiencing this intuitive, expressive and precise control redefined what professional creators expected from digital tools, influencing the design of creative software features and interactions.

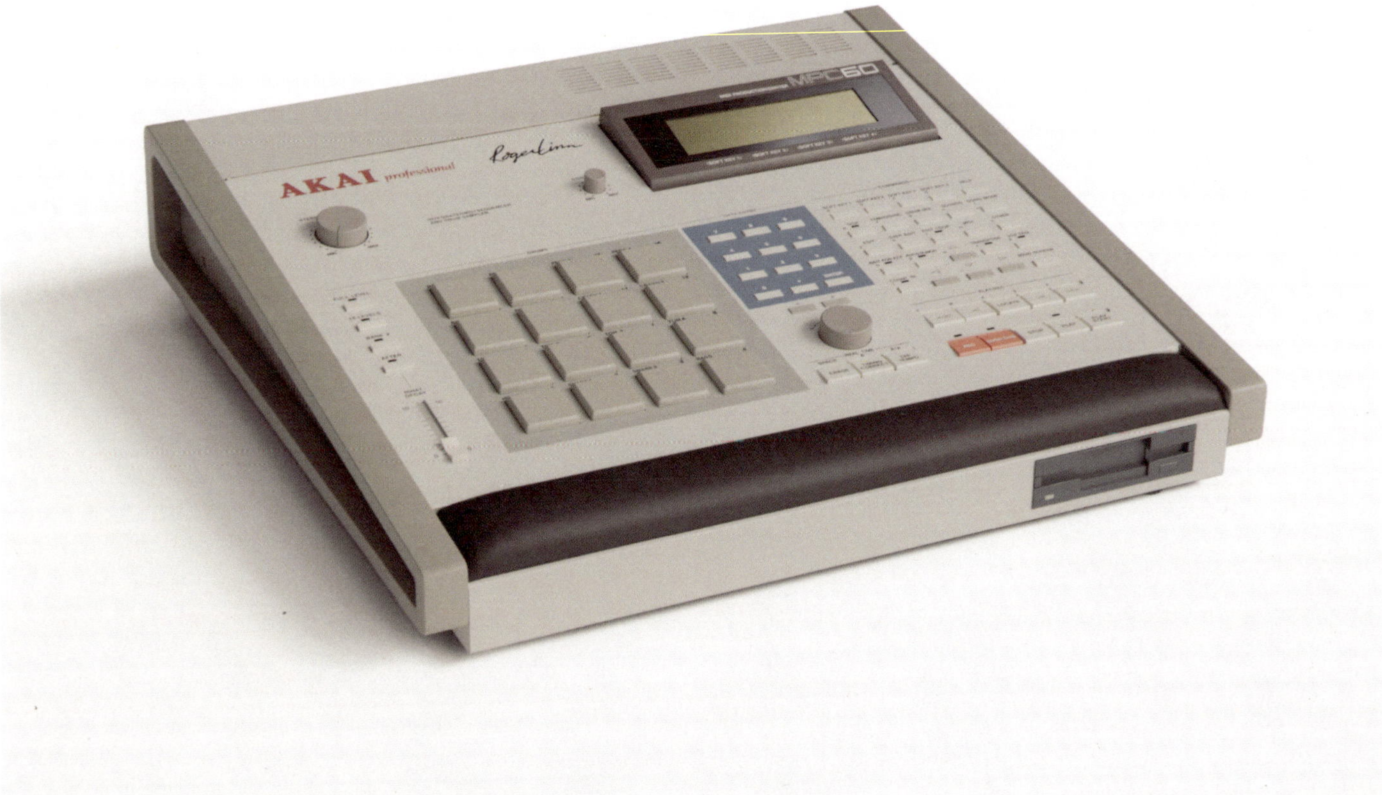

37 | Akai MPC Series

Category: Music

Release: 1988

Designer: Roger Linn

Manufacturer: Akai, Japan
(now Akai Professional, part of inMusic)

Dimensions: approximately
495 mm × 127 mm × 471 mm

Introduced as the MIDI Production Centre (later renamed Music Production Centre), the MPC60 was the first device to combine digital sampling and sequencing into a single, pad-based workstation. It allowed musicians to record, play, and arrange sound clips with expressive and rhythmic nuance.

Conceived by Roger Linn, the designer behind the pioneering LM-1 and LinnDrum, the MPC embodied an ambition to create a machine that behaved less like a piece of technical equipment and more like an instrument. Beyond drum samples, producers (many of whom did not play traditional instruments) learned to slice, loop, and overdub sounds and song fragments, crafting rhythms and building compositions that helped define the lean, punchy aesthetic of 1990s hip-hop and dance music.

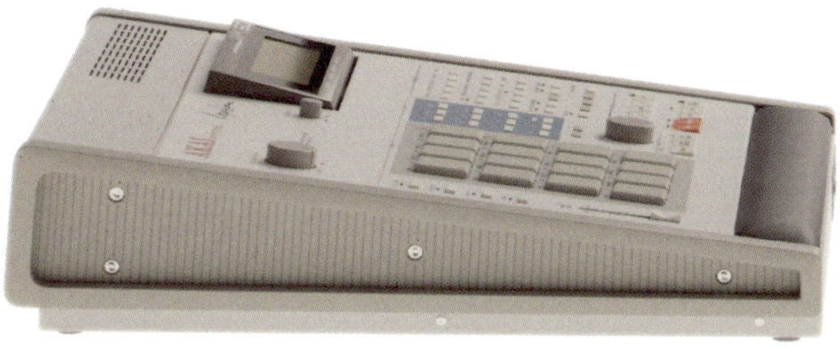

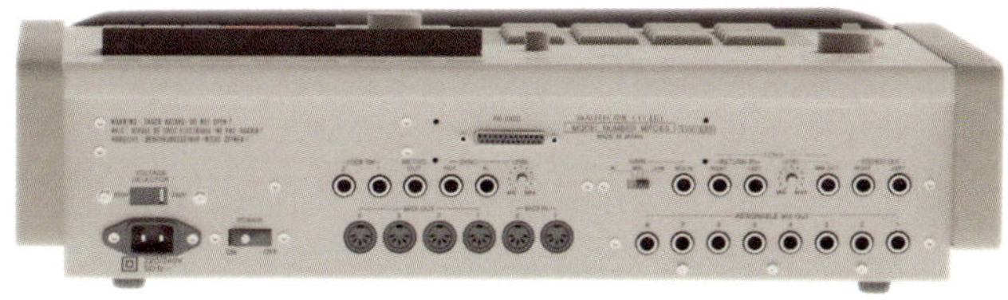

Design & Features

The MPC is self-contained in a single, robust, portable unit. 16 large, velocity- and pressure-sensitive pads respond with the expressiveness of a drummer or keyboardist, establishing a new standard for digital instrument control and performance in both studio and live settings.

Building on Linn's earlier work with the LM-1 and LinnDrum, the 'Swing' function lets users subtly shift the timing of alternate beats, infusing sequences with a natural, humanised groove.

Typical Workflow

1. Sample sound clips through the MPC's input: use a microphone or sounds and song parts from vinyl records. With limited sample memory, make every choice deliberate.
2. Use the edit screen (S) to trim, normalise, and loop each sound, assigning it to one of the 16 pads (P) to create a playable set.
3. Hit the pads to create percussive and melodic patterns. Keep a natural feel, tighten up the timing with quantisation, or add rhythmic nuance using 'Swing'. Build full arrangements by copying, looping, and overdubbing tracks, chaining sequences together to construct complete songs.
4. Ready to perform live? Mute tracks, switch sequences, and adjust sound levels on the fly, using the pads, buttons and data wheel.

The control is immediate, intuitive, and always within reach, reflecting a philosophy of keeping creative control front and centre.

Thoughtful user experience features include a dedicated help button, clearly labelled commands, a bright LCD, and tactile, hands-on controls. Together, these features make the MPC60 easy to operate, allowing musicians to focus on making music and exploring creative possibilities, rather than spending time reading the manual.

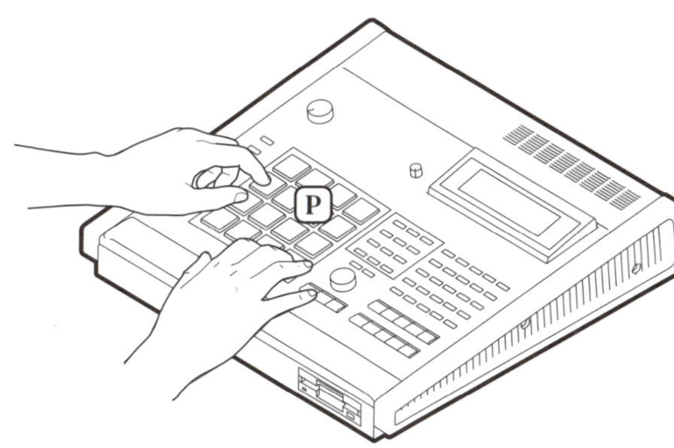

38 | Adobe Photoshop

Category: Photography, Illustration, Design

Release: 1990

Designers: Thomas Knoll, John Knoll

Manufacturer: Adobe, North America

Dimensions: N/A (128,000 lines of computer code when released)

When Thomas Knoll, a PhD student in computer science at the University of Michigan, created a small software application called Display to show grayscale images, his brother John Knoll, a visual effects artist, immediately saw its potential. Together, they developed the software into a powerful image editor, and eventually licensed it to Adobe; the commercial version of Photoshop was born.

Before Photoshop, digital retouching required expensive, specialised computer hardware. Its commercial release changed everything, bringing professional-grade image manipulation to the desktop, an innovation that had a lasting impact on photography practice and one that also reshaped the cultural perception of photographs. 'Photoshop' soon became a commonly used verb, shorthand for digitally altering images.

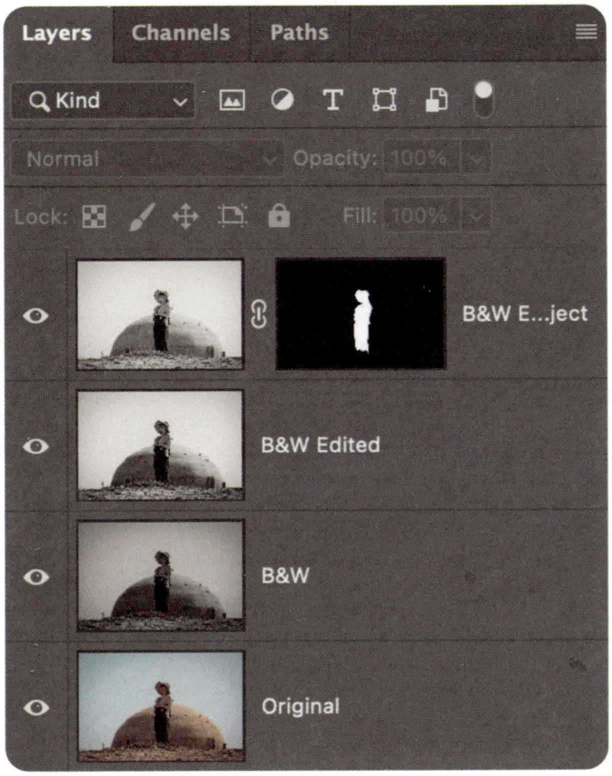

Design & Features

Photoshop's interface is organised around familiar physical creative concepts and processes: drawing, painting, and darkroom techniques in photography. Digital pens and brushes provide both expressive and precise control, while filters adjust colour, sharpen, blur, or stylise the image. Layers (introduced in 1994) work on different elements of an image independently, mirroring the mental models of collage, overlays, and stencils. Adjustment Layers enable non-destructive editing, without the fear of mistakes or the cost of redoing work.

The native file format (PSD) preserves edits as metadata and is widely used for moving images between Adobe and other applications. Lately, Photoshop has incorporated machine learning with tools such as Generative Fill and Sky Replacement, where users specify an intent and the software performs the technical work.

Typical Workflow

1. Offload high-resolution RAW images from the camera memory card (C). These contain the maximum detail and dynamic range, providing latitude for precise adjustments.
2. Use the Spot Healing Brush and Clone Stamp tools to remove imperfections by painting out dust and distractions.
3. Apply Adjustment Layers (A) to modify exposure, contrast, and colour balance, keeping the original pixels untouched.
4. Add a Layer (L) for creative edits: blend and composite multiple images, or apply visual effects. Create a Mask (M) to apply changes to selective areas of the image.
5. Export the finished work *flattened* (for example, as a JPEG, TIFF, or PNG), and also save a PSD version for future revisions.

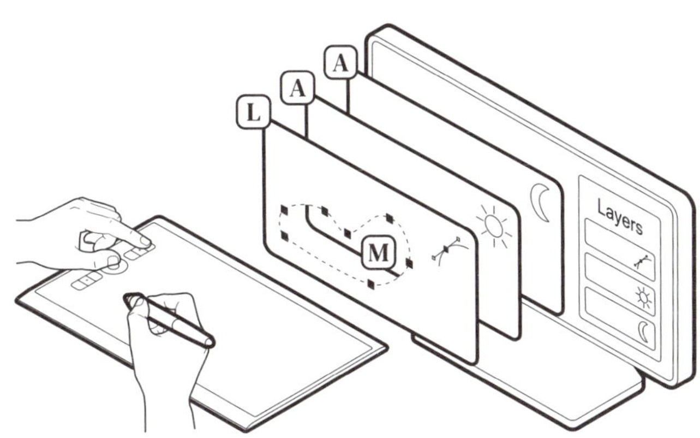

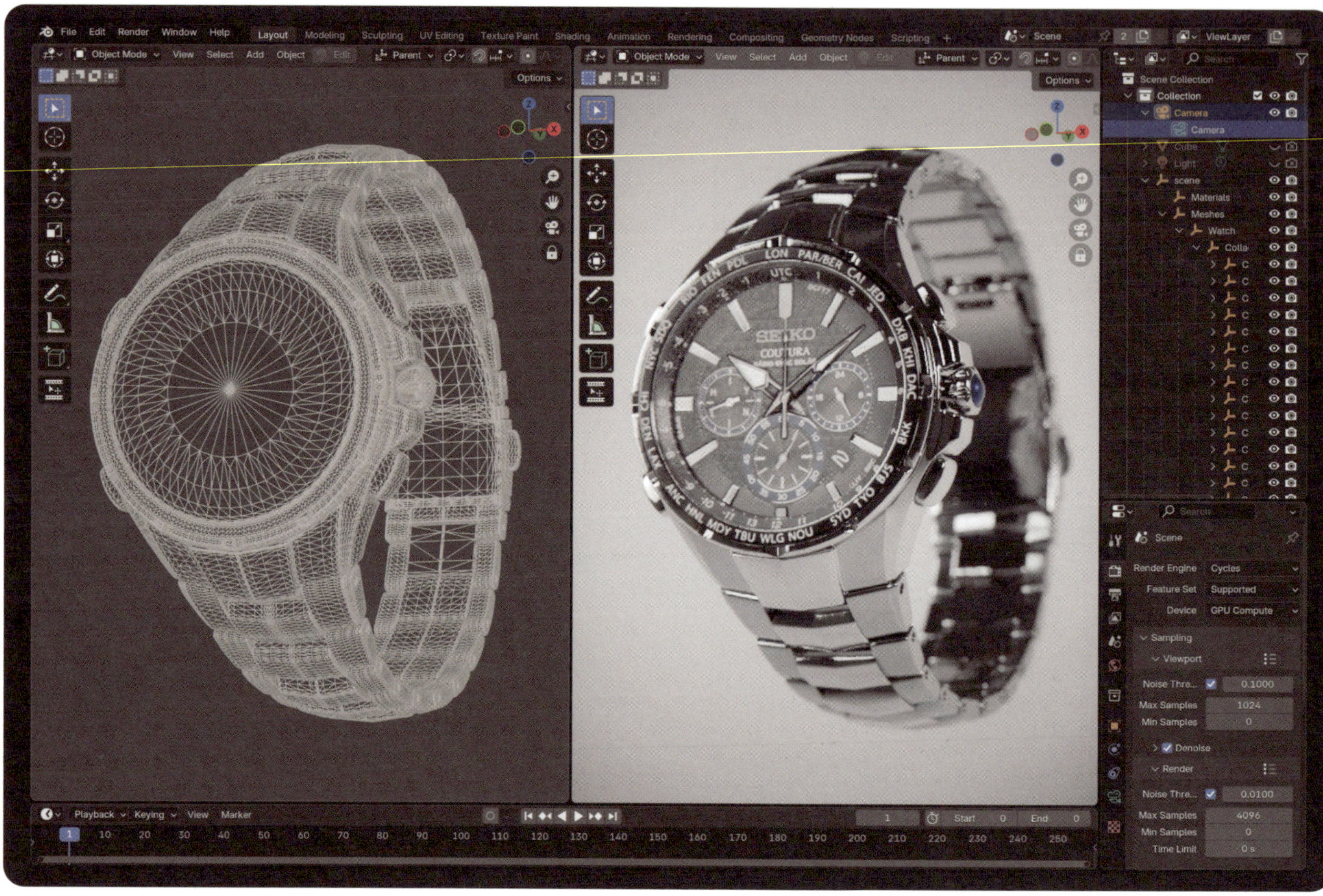

39 | Blender

Category: Design, Animation, Film

Release: 1995, 2002 (open source)

Designer: Ton Roosendaal

Manufacturer: Blender Foundation, the Netherlands

Dimensions: N/A (millions of lines of computer code)

Blender is a versatile 3D creation software suite used in animation, visual effects, game development, digital art, and visualisation. Developed initially by Ton Roosendaal as an in-house tool for his Dutch animation studio, it was designed to make 3D production faster, more flexible, and less painful, particularly amid clients' constant revisions.

In 2002, Roosendaal established the Blender Foundation and released the software as free, open-source technology. Since then, Blender has evolved into a full 3D production environment for modelling, sculpting, animation, rendering, and compositing. Ongoing development is driven by a global community of artists and developers, bringing together professional-grade capability with an open, collaborative ethos that champions accessibility and innovation.

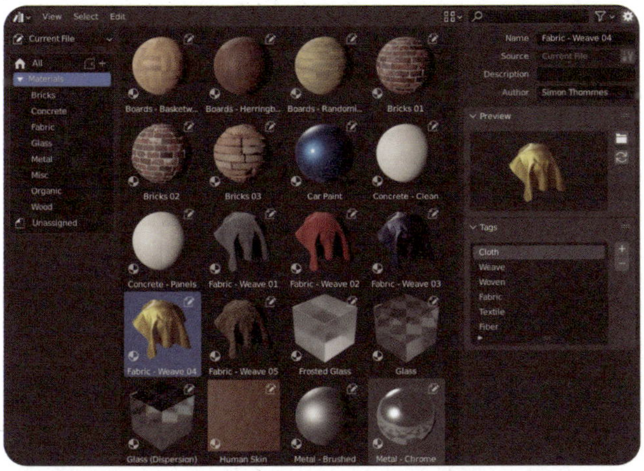

Design & Features

Blender unifies every stage of 3D production in a single environment, removing the friction of switching between separate applications. Its modular, customisable interface can be tailored to the needs of individual workflows. Innovative tools like Grease Pencil allow artists to draw in 2D, within a 3D space, enabling hybrid styles that blend traditional and digital animation.

Blender's design is shaped by equal parts technical ingenuity and community-led evolution. The open-source licence provides complete access to its underlying code, revealing not only how the software works but also inviting anyone to shape its future. This transparency, combined with a steady flow of ideas, fixes, and new features from contributors, has produced a tool that evolves quickly and embodies the needs of a diverse creative community.

Typical Workflow

1. Begin with a concept or sketch and translate it into three dimensions using Blender's modelling, sculpting, and layout tools.
2. In the Shader Editor, build a visual look, selecting colours, textures, and materials that mimic metal, glass, fabric, and more.
3. Bring the scene to life with the Animation Toolkit: rig models with mechanics and bones, set keyframes, and choreograph motion.
4. Place a camera to define the viewer's perspective and set the mood with lighting. Render as a still image or animated movie.
5. Complete your work with Blender's built-in post-production environment: composite layers, refine colour, and create a timeline edit, without leaving the software.

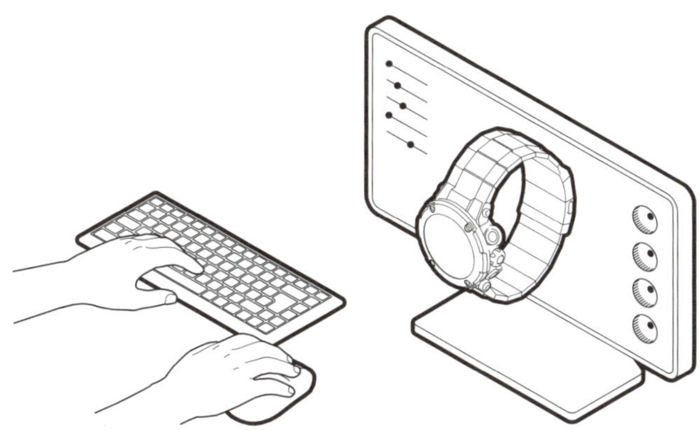

40 | Unreal Engine

Category: Video Games, Animation

Release: 1998

Designer: Tim Sweeney

Manufacturer: Epic Games, North America

Dimensions: N/A (millions of lines of computer code)

Unreal Engine started in the mid-1990s as a software technology project by Tim Sweeney, founder of Epic Games. Originally developed to power the first-person 3D shooter Unreal (released in 1998), the engine's broader purpose was well defined from the start: a flexible, extensible tool that could evolve across generations of games and be distributed across multiple platforms.

Today, Unreal Engine represents one of the most advanced universal real-time creative platforms available. It combines industry-leading graphics rendering, comprehensive world-building tools, a powerful visual scripting system, and full source code access, bringing interactive experiences to life across video games, film and animation production, architectural visualisation, automotive design, simulation and more.

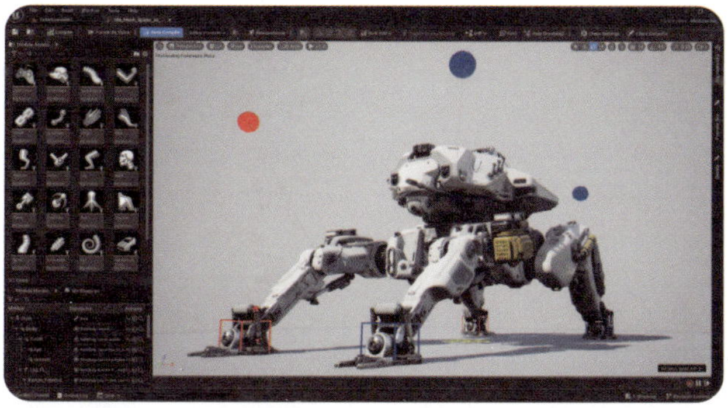

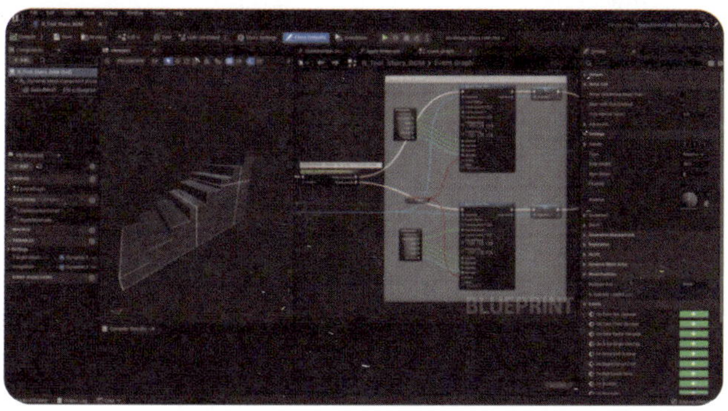

Design & Features

Unreal Engine is built for visual, iterative, and collaborative creation, an approach that aligns with how modern creative teams think and work. At the heart is the Unreal Editor, a unified workspace where artists, designers, and programmers develop assets, construct virtual worlds, define behaviours, preview gameplay, and refine ideas. Because art and development teams work simultaneously within the same real-time, modular environment, the gap between early concept exploration and fully playable results is dramatically reduced.

The Blueprint system offers high-level programming through visual, node-based logic, removing the traditional divide between design goals and optimised code. Plug-ins and full source code access enable extensive customisation, community contributions, and the ongoing evolution of the engine itself.

Typical Workflow

1. Select from a range of templates with built-in controllers and basic game structures.
2. Use the Level Editor to construct a playable world. Import existing assets or create new ones: 3D models, textures, animations, and sounds. Use parametric tools to quickly generate, populate, and style large scenes.
3. Define interactions and gameplay, building Blueprints to control actors, trigger events, and define player and AI responses.
4. Try playing directly in the Editor and test changes interactively in real time. Iterate and improve the design.
5. Package the project. This will automatically optimise and publish your game to desktop, mobile, console, and immersive platforms.

Akai MPC Series Cultural Impact

Roger Linn's 16-pad, sample-playing MPC design enabled producers to perform recorded sound clips with the physical expressiveness of a drummer or pianist. For many of hip-hop's foundational figures, it became their primary creative tool, with an influence so profound that the machine itself came to symbolise an entire era of hip-hop production.

By offering a compact, affordable alternative to a full professional studio, independent artists were able to transform fragments of existing records into new cultural dialogues. James Yancey, known as J Dilla, became one of the most influential beatmakers of the 1990s. After working with analog tape decks and early digital samplers, he found his voice on the MPC60 and later the MPC3000, which he used for most of his work and is now housed at the Smithsonian National Museum of African American History and Culture. What set Dilla apart were his hand-played rhythms, the MPC allowing for slight timing shifts that gave his beats a loose, human feel. He brought swing and emotional nuance into electronic production and permanently changed how producers think about groove. In the early 2000s, Kanye West helped reshape mainstream hip-hop by using the MPC2000 as a compositional instrument, chopping soul and gospel songs into short phrases. Treating the MPC as a melodic instrument, he expanded the hip-hop producer's creative songwriting role, using restrained drums to let the sampled chords and vocals carry the emotional weight of his songs.

Together, these musicians demonstrated how the MPC could redefine hip-hop as a producer-led, sample-based, and musical art form, driven by careful selection of sampled sources that conveyed personal style.

Adobe Photoshop Cultural Impact

Synonymous with image editing and emblematic of a broader shift in visual culture, Photoshop introduced new aesthetics while redefining expectations about what images could be and how they might be manipulated. People talk about 'Photoshopping' just as they might say 'Googling' a topic, which reflects how deeply the tool has entered everyday practice.

Andreas Gursky used Photoshop to construct monumental, hyperreal scenes, stitching multiple photographs together and erasing distracting details. His digitally composed works challenged the assumption that photographs must be strictly documentary, helping to legitimise digital manipulation within fine art. David LaChapelle similarly embraced Photoshop to amplify colour, scale, and theatricality, producing saturated, surreal images that defined the visual language of late-1990s and early-2000s pop photography. In his hands, Photoshop became a narrative device rather than a corrective tool. Annie Leibovitz integrated digital retouching into her largely traditional portrait workflow. In publications such as Vanity Fair, Photoshop enabled a heightened sense of perfection and control, subtly blurring the boundary between photography and illustration within mainstream editorial culture.

However, these achievements brought significant consequences. The perfection of human features in the media created unrealistic beauty standards and debates about healthy body image. The notion that photographs are no longer objective records prompted many to question the authenticity of images and the truth of journalism. How we perceive reality has changed, moving culture into an era in which every image is viewed with a degree of scepticism.

Blender Cultural Impact

Blender has evolved from a niche open-source project into a tool that fundamentally changes access to 3D modelling and animation, removing the cost barrier of using commercial software. Its cultural impact comes from a radical commitment to transparency and community ownership. Unlike commercial tools that guard access and their secrets behind paid subscriptions, Blender is freely accessible to students in schools, enthusiasts in bedrooms, and independent teams of professional creators everywhere.

Its influence continues through Open Movies: animated short films produced by the Blender Studio, in which all production assets (models, textures, rigs, and animation files) are released under a Creative Commons licence, allowing anyone to download, study, and reuse them in their own work. This treats the filmmaking process itself as a platform for learning and development, for both the Blender development team and its users, helping strengthen an open culture in which sharing and remixing are integral to artistic practice. Technical growth and new features can be guided by production experience rather than hypothetical ideas, with the sharing of project data and an open production pipeline. Specific design decisions have sparked unexpected cultural trends. The Grease Pencil tool (intended initially for simple 2D annotations during 3D production) was repurposed by the community into a hybrid system for mixing 2D and 3D animation, resulting in a distinctive visual style that combines structure and depth with the painterly impression of hand-drawn art.

Beyond the images it helps create, Blender's impact has been far greater due to the values it embodies: openness, experimentation, sharing, and collective authorship.

Unreal Engine Cultural Impact

The video game Unreal (released in 1998) and its multiplayer sequel, Unreal Tournament (1999), marked a pivotal moment when the game engine shifted from a hidden, technical foundation into a prominent creative tool. While earlier titles such as Doom established the FPS (First-Person Shooter) genre, Unreal integrated a map editor (Unreal Editor) and programming language (UnrealScript). This provided a full suite of development tools to players at no additional cost and enabled a large community to learn game development skills, reinforcing Unreal as an engine, a training ground and a pathway into game development.

By treating players as potential developers, the Unreal ecosystem broke down the wall between consumers and creators. It establishes a culture in which the creative tool itself can be as iconic as the work it produces. Unreal Engine (UE) evolved through a combination of technical ambition, openness to users, and strategic expansion beyond games. UE2 (2001) and UE3 (2004) expanded the game engine's capabilities to include high-performance real-time 3D graphics and middleware features such as physics, animation systems, and, later, visual scripting, making the engine attractive to large studios by reducing their need to develop proprietary technology. A key shift occurred when UE4 (2014) offered free developer access, with no cost to a developer until their games were released and generated significant revenue.

UE5 (2022) is used across video games, film and animation, architecture, automotive design, and simulation. Its growth reflects a shift in creative culture toward ubiquitous real-time 3D workflows and the impact of lowering barriers to entry to professional-level creative tools, making them accessible to individuals and small teams, with no upfront costs.

41 | Pioneer CDJ Series

Category: Music

Release: 2001

Designer: Pioneer

Manufacturer: Pioneer DJ, Japan (now AlphaTheta Corporation)

Dimensions: approximately 320 mm × 110 mm × 370 mm

The Pioneer CDJ-1000 digitised DJ performance, bringing the tactile feel of playing vinyl records to the popular CD format. A large, responsive jog wheel allowed DJs to cue, scratch, and pitch-bend with the same expressive control as an analog turntable, while offering the precision, reliability, and convenience of digital playback.

It was quickly adopted by clubs and festivals worldwide and became the professional standard. DJs could carry their music on CDs rather than in heavy crates of records, while retaining their hands-on artistry of live mixing. The tool also enabled new creative gestures such as seamless looping, instant hot cues, and the ability to change a track's tempo without altering its musical key, all of which expanded the DJ vocabulary and cemented the CDJ as an iconic instrument in club culture.

Design & Features

Drawing on Pioneer's experience building reliable CD players for cars, the CDJ features a robust construction that withstands the demands of high-energy club environments. Its modular form and horizontal layout echo the familiar design of a Technics turntable. The large pressure-sensitive jog wheel replicates the resistance of a turntable platter, providing the tactile control DJs rely on while adding the portability and flexibility of digital. A bright interface clearly displays cue points and live audio waveforms, coupled to dedicated buttons for precise cueing and flexible looping.

Time once spent digging through record crates and finding a cue point by carefully dropping a turntable needle can now be used to construct and layer tracks, live-mix, and apply built-in effects, opening up new ways to build and drive the creative performance.

Typical Workflow

1. Find a CD to start your set, load it into the CDJ and select a track.
2. Spin the jog wheel (J) while scanning the waveform display (D) to find song sections. Tap CUE (C) to mark precise locations.
3. Press PLAY (P) to listen. Slide the TEMPO fader (T) to match the pitch/tempo of another track playing on a second CDJ.
4. Use dedicated buttons to trigger hot cues, loop, extend, and reshape the performance. Tap the jog wheel to make minor timing adjustments, and 'scratch' to add style and engage the audience.
5. Blend between tracks over an extended mix, allowing the jog wheel to slow under your hand. Cut sharply to the next cue point for a powerful, abrupt transition.

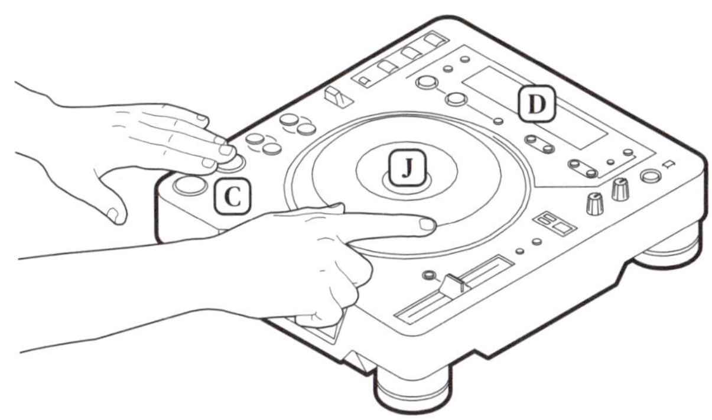

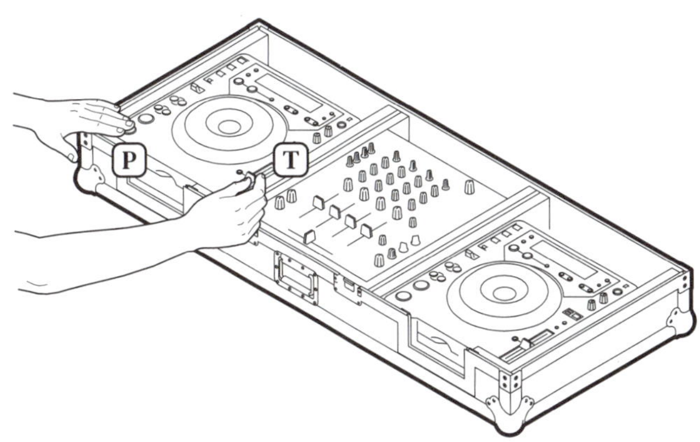

42 | Canon EOS 5D Mark II

Category: Photography, Video

Release: 2008

Designer: Canon

Manufacturer: Canon, Japan

Dimensions: approximately
152 mm × 114 mm × 75 mm

The Canon EOS 5D Mark II was the first full-frame DSLR camera to integrate high-definition video recording. Its significance lay in introducing a cinematic aesthetic to digital video, achieved through a range of interchangeable photographic lenses with shallow depth of field, distinctive bokeh, and low-light performance.

A compact form factor and relative affordability put these attributes within reach for independent filmmakers, photojournalists, videographers, and online bloggers, sparking a significant cultural shift. By enabling a new generation of creators to emulate the look of professional film production, the 5D Mark II laid the foundation that ultimately contributed to the rise of the modern digital influencer and content creator ecosystem.

Design & Features

The unassuming camera body departs from the typical video camera shape. Familiar ergonomics and having the physical controls and layout of a stills camera allows photographers to transfer their existing skills directly to filmmaking. Working with a compact and familiar profile also allows filming without attracting attention, giving documentary makers and journalists a discreet advantage in the field.

Compatibility with Canon's extensive EF lens catalogue provides immediate access to a broad range of visual styles. Wide apertures offer shallow depth of field and distinctive bokeh, while fast prime lenses make low-light shooting achievable without the bulk or expense of dedicated equipment. The rear LCD screen with Live View frees the filmmaker from a traditional shoulder-mounted viewfinder, enabling more intimate, responsive framing.

Typical Workflow

1. Mount a Canon EF lens onto the DSLR body.
2. Set shutter speed, aperture, and ISO using the familiar physical dials (D), noting that shutter speed is now tied to the video frame rate and will affect motion blur. A guideline is to set the shutter speed to roughly twice the frame rate to achieve a natural, cinematic appearance.
3. Activate live view on the rear LCD screen (S) to compose the frame and check focus.
4. Press the shutter button (now serving as a record button) (R) to begin capturing high-definition video to the internal CompactFlash memory card.
5. Play and review footage on the LCD screen. Transfer the media files to a computer using the memory card or a USB cable.

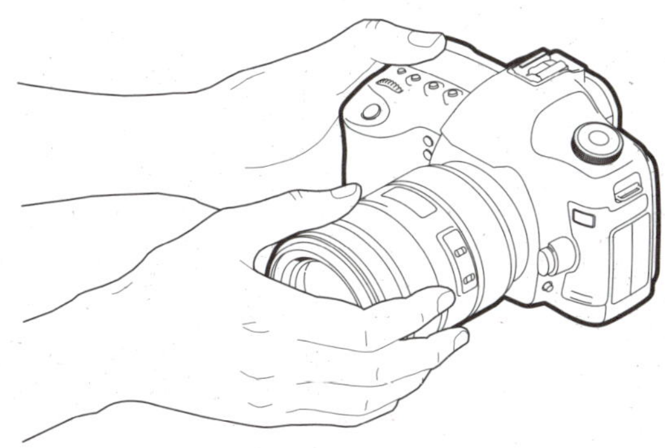

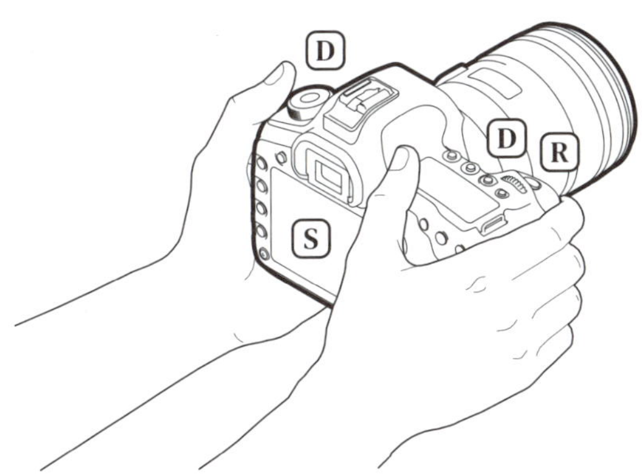

▲ *DJ Playing Music*

Pioneer CDJ Series Cultural Impact

The CDJ successfully bridged a tradition of club culture with the rapidly emerging digital future. Closely emulating the feel of a vinyl turntable, it introduced a familiar interface and the reliability and creative control of digital media. For established DJs, this balance of continuity with innovation was essential to their adoption.

House and techno DJ Carl Cox used the CDJ alongside his vinyl records, seeing it as a natural evolution. He highlighted the ability to play loops, samples, and rapid track changes, impossible on traditional turntables. Cox was wary of laptop-based systems, which he felt distanced performers from the physical act of DJing and introduced risks, such as unreliable audio and computer crashes. Both were addressed by the CDJ's purpose-built hardware and dedicated embedded software. Dance and electronica DJ Norman Cook, known as Fatboy Slim, initially resisted the CDJ, having spent his career mastering the nuances possible with vinyl records. His perspective changed through practical experience: facing unreliable club turntables and a back-to-back performance with David Guetta on a moving lorry, where Guetta's use of the CDJ proved both louder and more reliable, prompting Cook to make the switch, humorously noting that he could now dance freely in the DJ booth without causing his tracks to skip.

Beyond performance, the CDJ changed the relationship between production and playback. Producers could burn CDs (and later use USB drives) to circulate unreleased tracks, shortening the gap between creating music and hearing it live on the dance floor. This accelerated the evolution of dance music in the 2000s, shaping club culture and establishing a new, digitally fluent era of DJing.

Canon EOS 5D Mark II Cultural Impact

Full-frame 1080p video recording built into the 5D Mark II was not originally intended as a professional tool. Still, it enabled, for the first time, cinematic imagery at consumer prices, giving independent filmmakers access to a look associated with high-end film production. An early example of this is Reverie (2008), a short film made by French-American director and Pulitzer Prize-winning photographer Vincent Laforet. He used a pre-production 5D Mark II lent to him by Canon over 72 hours, capturing handheld shots and sweeping cityscapes. The footage circulated widely and inspired many to reconsider which tools were necessary to produce cinematic work.

The 5D Mark II also proved to be a capable tool in mainstream production. Image quality and interchangeable lenses enabled its adoption in Hollywood as a 'crash' camera, placed in extreme locations to capture dynamic action. The episodic show House, M.D. shot a full episode using one, demonstrating that such a camera could now work for television. But the broader impact remained most noticeable within the independent and creator communities. Film and video makers without studio backing or high-end equipment were able to step up: wedding videography transformed from simple videos into cinematic storytelling; science, nature and history documentaries gained tools that were unobtrusive and produced compelling images in low light; and YouTube creators attracted audiences with good-looking videos.

The 5D Mark II had limitations: it lacked professional audio controls and exhibited image-warping rolling-shutter artefacts in fast-motion shots. How filmmakers developed workarounds and adaptations to these challenges shows that communities adapt new creative tools to work beyond their initial design.

43 | Novation Launchpad

Category: Music

Release: 2009

Designers: Novation, Ableton

Manufacturer: Novation Music, United Kingdom

Dimensions: approximately 239 mm × 24 mm × 239 mm

Developed in partnership with Novation and Ableton, Launchpad redefined the relationship between musician and digital audio software. It emerged at a moment when software interfaces dominated, threatening to decouple music-making from physical, tactile interaction. This new on-desk device translated Ableton Live's modular Session View into a tangible, light-filled interface. Clarity was core to its design: a simple grid of 64 illuminated pads that let users trigger, loop, and transform sound clips in real time.

The glowing grid evolved into a cultural emblem, a visual shorthand for electronic creativity in the YouTube creator era. It supported a convergence of precise studio production and expressive visual performance for a generation of music producers, DJs, and live performers.

Design & Features

The minimal yet highly adaptable 8x8 matrix of pads, framed by contextual controls, functions as both a direct interface and a dynamic visual display. Colour illumination provides clear visual feedback, and velocity- and pressure-sensitive pads support expressive multi-touch for performance-driven interaction. Bidirectional communication with Ableton Live updates pad colours and relays the current status of sound clips, tracks, rhythms, and effects. Each pad serves as both a control and an information marker, transforming the surface into a dynamic map of instruments and performance states.

Compact, durable, and powered directly via USB, the design is effective for both studio and live-stage use. MIDI connectivity inspires users to find applications beyond music, making it a popular interface for lighting designers and visual artists who need interactive control.

Typical Workflow

1. Connect Launchpad to Ableton Live using a USB cable. Assign sounds and music clips to the pad grid. Visual feedback reflects selections made in the software.
2. Hit the pads (P) to audition sounds, trigger loops, and build song arrangements.
3. Use the contextual controls (C) to switch modes, mix, and adjust levels, without returning to the computer screen or mouse.
4. Play the grid as a performance instrument. Trigger, mute, and reorganise musical elements and create coordinated light patterns by setting timed pad illumination.
5. Develop muscle memory and discover how a seemingly simple controller evolves into a personalised, expressive instrument.

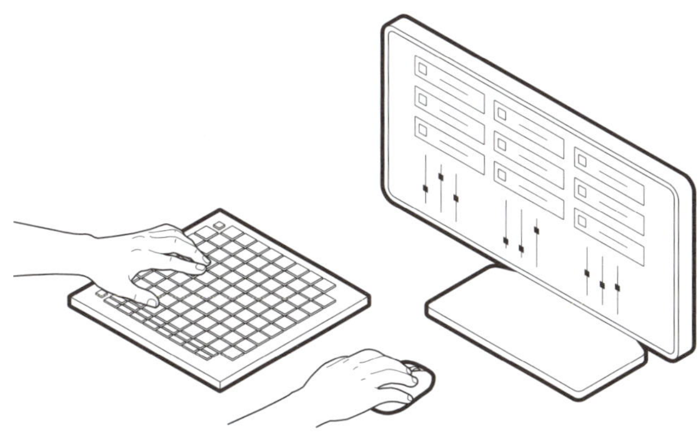

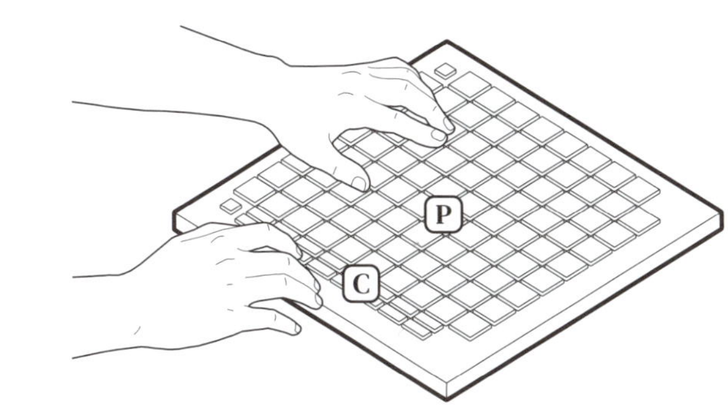

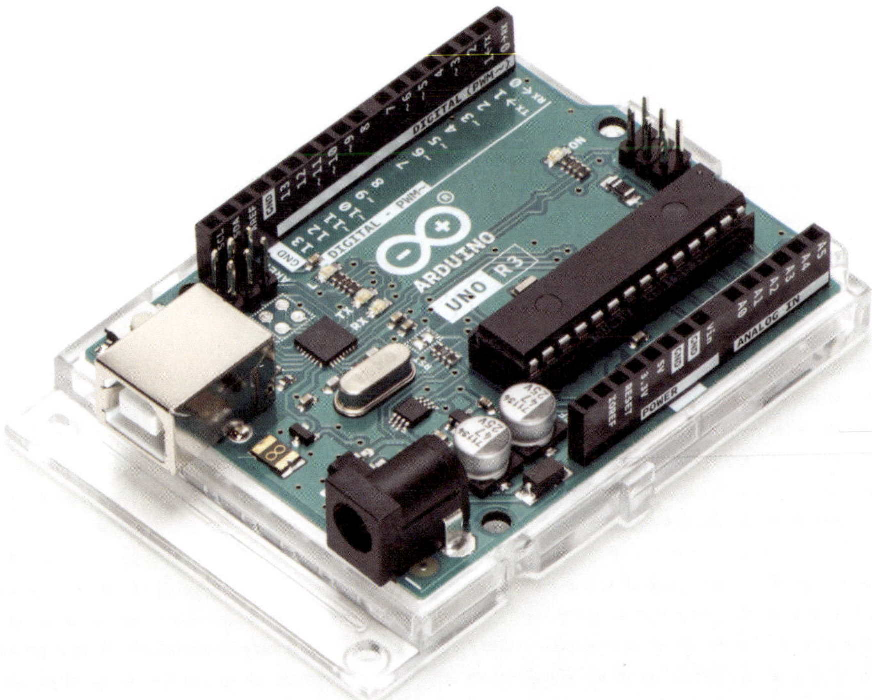

44 | Arduino UNO Board

Category: Design, Making

Release: 2010

Designers: Massimo Banzi, David Cuartielles, Tom Igoe, David Mellis

Manufacturer: Arduino, Italy

Dimensions: approximately 69 mm × 53 mm × 25 mm

Conceived in 2005 at the Interaction Design Institute Ivrea (IDII) in Italy, Arduino emerged as an accessible prototyping platform for interactive objects. It empowered artists, students, and makers to explore ideas without requiring a specialised engineering skillset. The board crystallised an ethos of openness, simplicity, and community. Its architecture allows designers to connect physical inputs (such as switches or sensors) to outputs (such as lights, speakers, motors, or displays), establishing rich interactions between the digital and physical worlds.

Arduino has evolved to transcend its role as a prototyping tool, becoming a movement that underpins hands-on education and everything from playful installations to the rapid development of full-scale commercial products.

Design & Features

The microcontroller board is designed for plug-and-play ease of use. Its aesthetic and function combine a logical layout, clear labelling, and the instantly recognisable blue PCB, signalling an approachable, hands-on identity. 14 digital I/O pins and six analog input pins provide flexible connections to an array of sensors, indicators, motors, actuators, and other devices. The standard USB port handles power, programming, and testing.

Compact, standardised, and lightweight, the form factor readily accommodates modular add-ons and stackable Shields, enabling expansion with features such as Wi-Fi networking. Both the hardware and software are open source. This ensures they are freely available for modification and remixing, fostering a dynamic ecosystem of learning and shared knowledge.

Typical Workflow

1. Prototype an idea (maybe a sound player!). Using jumper wires (J), connect a button (B) and speaker (S) to the IO pins (I).
2. On a computer, use the Arduino IDE (Integrated Development Environment) to write a control program that plays a sound when the button is pressed.
3. Upload the program to the board via USB to bring the circuit to life. Does it work?
4. Adjust the wiring or refine the program code. A cycle of rapid prototyping (experiment, observe, and adjust) drives the creative design process.
5. Try powering the board from a battery and test-fit it into a mock-up product. Ask questions and share your results with others in the online communities.

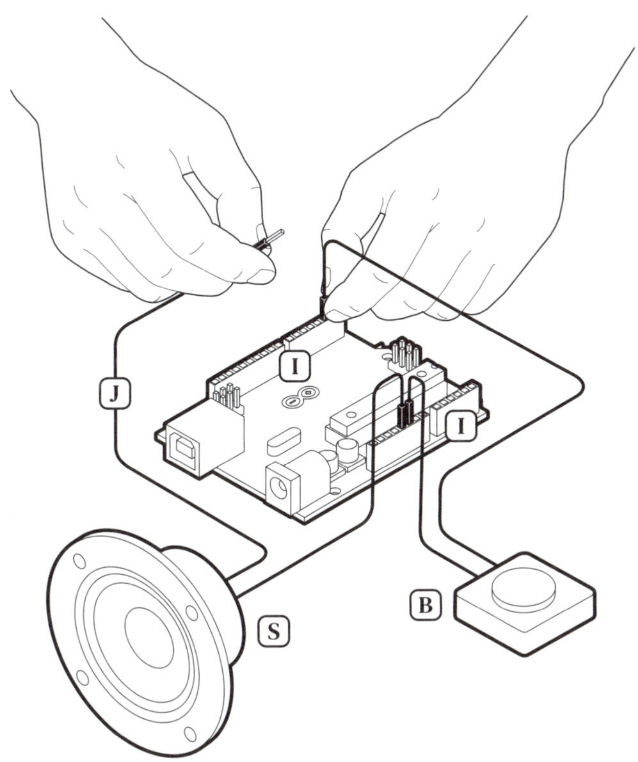

Novation Launchpad Cultural Impact

The Launchpad's intuitive, visually responsive grid layout opened electronic music creation to players, with and without traditional instrument training, allowing them to arrange and perform songs using a structured grid interface.

Its first significant cultural moment arrived in 2011 with Pop Culture, a performance by French musician and producer Hugo Pierre Leclercq, known as Madeon. At just 17 years old, and after only one day of using his newly purchased Launchpad, Madeon performed a real-time mash-up of clips from 39 songs. The video quickly went viral, introducing millions of viewers to his artistry and to Launchpad's expressive potential. It demonstrated that a producer could be as physically and visually engaging as a traditional instrumentalist, redefining how electronic music could be presented on video platforms such as YouTube. Producer and DJ Nick Boundy, known as M4SONIC, extended clip launching into a form of performance art, establishing a lightshow culture in which intricate, animated visual patterns were integrated closely with music composition. By touring alongside festival performances, he elevated Launchpad playing from a novelty into a recognised practice within EDM.

Launchpad contributed to the growth of *controllerism*, which emerged in the early 2000s as electronic musicians and DJs responded to the rise of laptop-based performance and digital DJing. As software made it easier to manipulate music in real time, performers began using dedicated hardware interfaces to interact with sound more physically and expressively. Influenced by turntablism's view of DJ gear as instruments, controllerism reframed digital tools as playable performance interfaces.

Arduino UNO Board Cultural Impact

Arduino's cultural impact reaches beyond its origins as a simple electronics board. As co-creator Massimo Banzi explained, it 'emerged from a desire to remove the need for permission to build something great'[11], a conviction that informed a tool that removed barriers to hardware development and transformed electronics into a participatory medium. Arduino directly contributed to the maker movement of the 2000s, fostering a culture of do-it-yourself innovation that blended electronics, computing, robotics, and craft. Use cases circulated through online forums, community workshops, and inventor fairs, where learning, sharing, and adaptation are celebrated. This ethos aligned with open-source values, encouraging playful *tinkering* and reducing the stigma associated with technology and electronics.

Rafael Lozano-Hemmer's Pulse Room used sensors to translate participants' heartbeats into a display of flickering incandescent light bulbs. At Disney Research, Botanicus Interacticus made living plants into touch- and proximity-sensitive interfaces that trigger sounds and digital effects. In classrooms, Arduino significantly impacted educational culture, changing how electronics, programming, and design are taught and who feels allowed to participate. It helped shift education away from theory toward hands-on learning, moving quickly from idea to working prototype, encouraging experimentation and reducing fear of failure. Teachers adopted Arduino not only in engineering courses but also in art, design, and media studies.

Born in a design school rather than an engineering lab, the Arduino microcontroller successfully achieved its objective of reshaping engineering-led hardware practice.

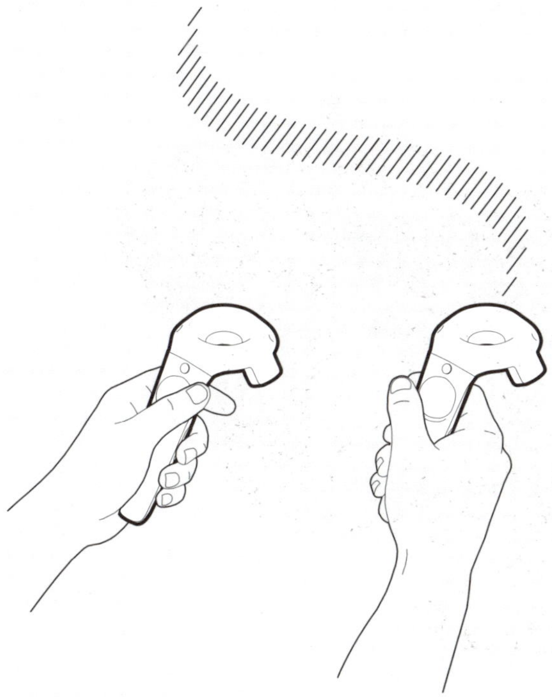

45 | Tilt Brush

2014 — Virtual Reality allowed digital artists and designers to step inside their work, painting immersively in three dimensions and turning the air around them into their canvas.

Tilt Brush tracks the user's head and hands, translating physical movement into paint strokes and volumetric forms that appear positioned within a room-scale space. The UI departs from traditional on-screen windows and menus, instead existing as an immersive experience within the space surrounding the user. Tools are accessed through simple, physical gestures: a palette floats, anchored to one hand, while brushes and colours are selected by pointing with the other, much like using a physical paintbrush on a real palette. For many users, this is their first encounter with an interface that closely aligns with natural movement, real-world intuition, and human perception.

46 | Figma

2016 — By reducing friction in the UI and UX design process, Figma enabled designers and clients to collaborate in the same space, share real-time feedback, and maintain shared design systems as a single source of truth. It was built on the principle that design software should live on the web, unlike traditional tools that relied on installed applications and manual file transfers. Figma runs directly in the browser, making work instantly accessible and easy to share.

Features such as component-based design, auto-layout, and real-time multi-user editing have changed how designers approach interdisciplinary collaboration, scalability, and consistency. As a result, Figma helped establish accessibility and collaboration as baseline expectations for modern creative tools.

► *Figma UI Component*

47 | ARRI ALEXA

Category: Film

Release: 2010

Designers: James Orrom, ARRI

Manufacturer: Arnold & Richter Cine Technik, Germany (ARRI)

Dimensions: approximately 321 mm × 224 mm × 153 mm (camera body)

When it debuted in 2010, the ARRI ALEXA marked a turning point in feature-film production, among the first digital cameras to convincingly stand shoulder to shoulder with 35mm celluloid film. Its combination of a Super-35mm CMOS image sensor, robust engineering, intuitive controls, and a streamlined file-based workflow enabled productions to shoot digitally without sacrificing the image quality, colour latitude, and confidence long associated with film.

Embraced across the industry, the ALEXA became a defining camera of its generation. From independent films and prime-time commercials to major Hollywood blockbusters, it set the template for digital cinematography in the decade that followed.

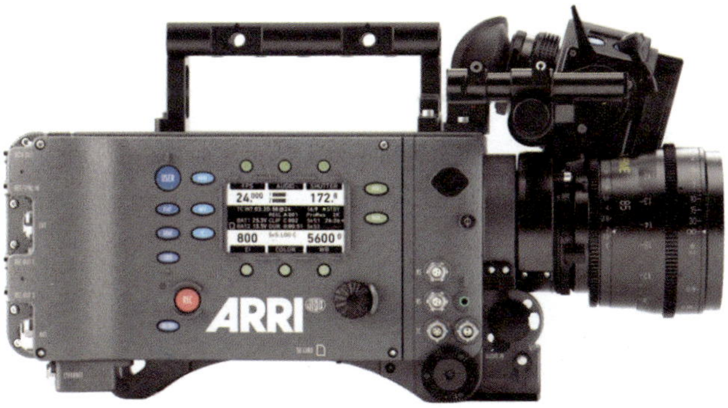

Design & Features

Designed to feel familiar to film crews moving from 35mm film to digital, the ALEXA echoes the form of a traditional film camera, with balanced weight, sturdy handles, and mounting points for accessories, rigs, and shoulder use. The interface centres on a small screen with six buttons and a rotary encoder, giving direct access to menus without several layers of sub-navigation. The clarity and speed of this design proved so effective that other companies widely copied it.

By recording high-quality ProRes files directly to easy-to-handle memory cards, the camera removes many of the technical hurdles of earlier digital systems. ARRI's 'Log-C' imaging science reproduces the entire sensor range as a *digital negative* image. The shoot-to-post process feels direct and dependable, and more efficient than shooting on and processing chemical film.

Typical Workflow

1. Set up the camera body, an optical or electronic viewfinder (V), lens (L), accessories and rig for the shoot.
2. Use the menu screen (S), buttons (B), and rotary encoder (E) to set the recording format, white balance, and exposure, monitoring available latitude on the false-colour feedback display.
3. Frame your subject and adjust focus.
4. Action! Press REC (R) to capture directly to the internal memory card in ProRes format, or to an external data recorder in fully uncompressed ARRIRAW format.
5. Review playback on the camera or via an external monitor to confirm exposure, focus, and performance. Copy the media to create backups and mark clips for the edit.

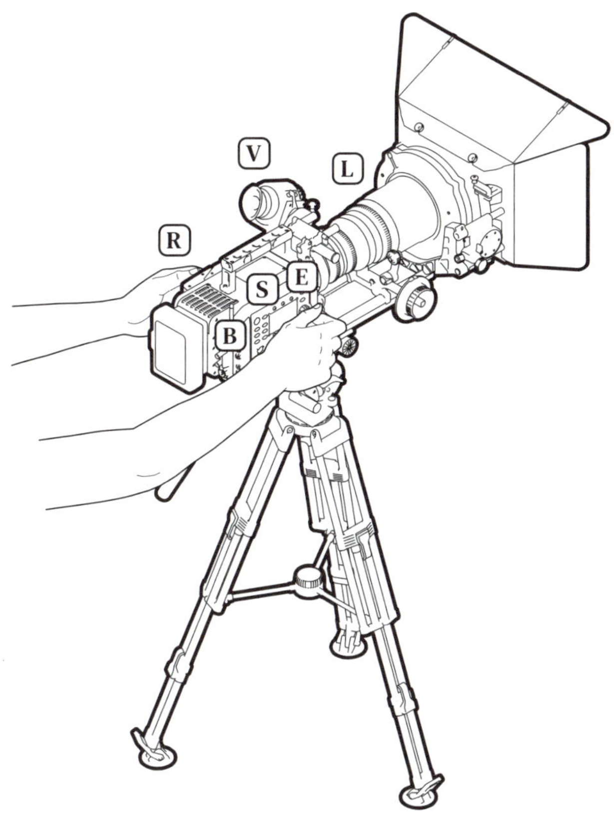

48 | Teenage Engineering OP-1

Category: Music

Release: 2011, 2022 (OP-1 field)

Designers: Jesper Kouthoofd, David Eriksson, Jens Rudberg, David Möllerstedt

Manufacturer: Teenage Engineering, Sweden

Dimensions: approximately 282 mm × 14 mm × 102 mm

When it arrived, the OP-1 overturned expectations about what a synthesiser and battery-powered portable studio could be. When most music hardware was bulky, utilitarian, and visually restrained, the OP-1 appeared as a slender, lightweight object with a bright, toy-like interface and a mix of Scandinavian and sci-fi design language.

It combined a synthesiser, sampler, four-track tape-style recorder, and FM radio in a device small enough to hold in one hand or carry in a backpack. The appeal lay in both technical versatility and a distinctive attitude: a blend of playfulness and design confidence that invited curiosity and experimentation over strict perfection. It remains one of the few modern digital instruments that functioned as both an object of desire and a genuinely expressive creative tool.

Design & Features

The OP-1 design embraces limitation as a guiding principle where fewer features encourage immediacy and intention, and the size and form invite explorative interaction.

Just four colour-coded rotary knobs map directly to on-screen parameters, changing with the active selection or mode. Each function is visualised through playful, characterful graphics on a bright OLED display, with instruments and effects shown as small, animated illustrations (a boxer for a punch filter, a cow for delay), making sound design fun and engaging. Mechanical accessories such as handles and rubber bands can be attached to the knobs, and an internal motion sensor enables sound to be shaped simply by tilting the device. The recording workflow draws on the logic of classic tape machines: overdubbing and splicing behave as they would on real magnetic tape.

A long-lasting rechargeable battery, together with the built-in speaker, microphone, and radio tuner, makes the OP-1 completely self-contained. It is at home in the studio or in more unexpected settings like a park bench, an aeroplane seat, or on top of a mountain, finding your musical inspiration!

Typical Workflow

1. Power on and start playing straight away. Choose a sound engine or sample from the built-in microphone (M) or radio receiver.
2. Let the animated visual metaphors (V) guide exploration as you shape sound parameters with the colour-coded dials (D).
3. Key in patterns with the step sequencer, or play live and record to 'tape'. Overdubbing is destructive; treat it as a living surface rather than a safety net. Finish your composition using compression and mix controls to add warmth and cohesion.

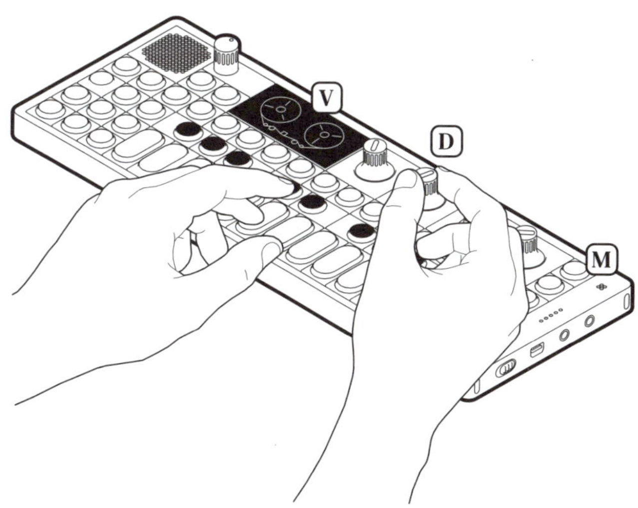

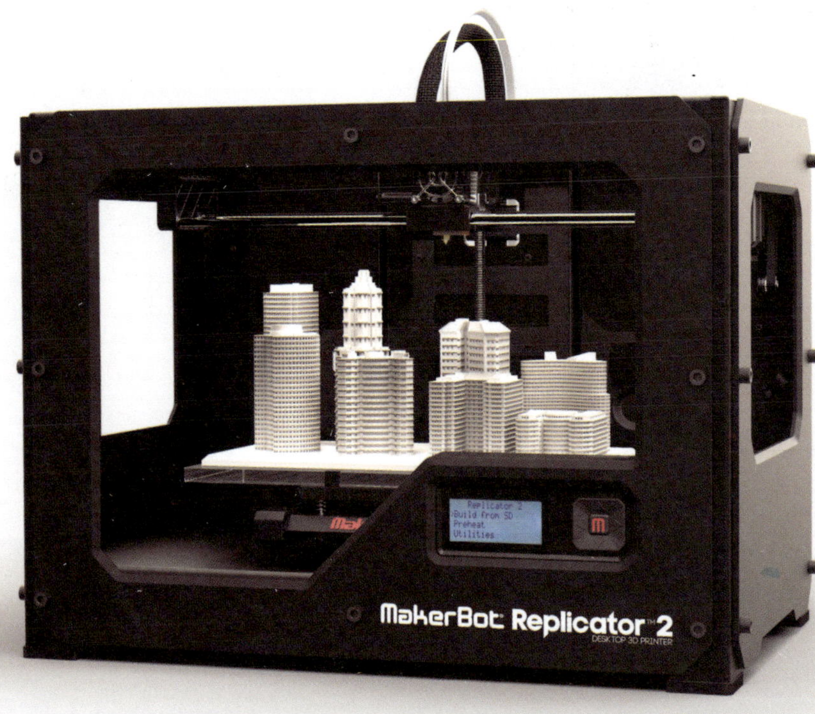

49 | MakerBot Replicator 2

Category: Making, Design

Release: 2012

Designer: MakerBot

Manufacturer: MakerBot, North America

Dimensions: approximately
490 mm × 380 mm × 320 mm
(without filament spool)

As one of the first machines to deliver professional-grade 3D printing to the desktop, the Replicator 2 redefined expectations for ease of use, print quality, and reliability. Conceived as an affordable, fast, and capable manufacturing tool, it opened a new chapter in the accessibility and affordability of small-part fabrication. By bringing rapid prototyping into classrooms, design studios, and small workshops, it enabled educators, designers, engineers, and a wider creative community to design, iterate, and produce their own plastic objects.

The Replicator 2 became a gateway to hands-on making and also a catalyst for engagement in a complete STEAM (Science, Technology, Engineering, Arts, and Maths) learning ecosystem.

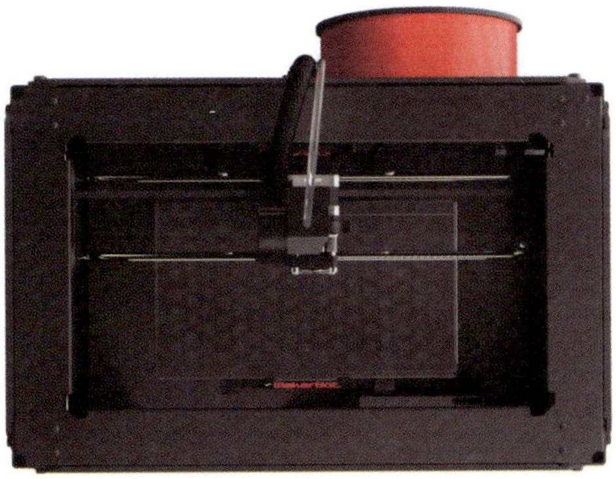

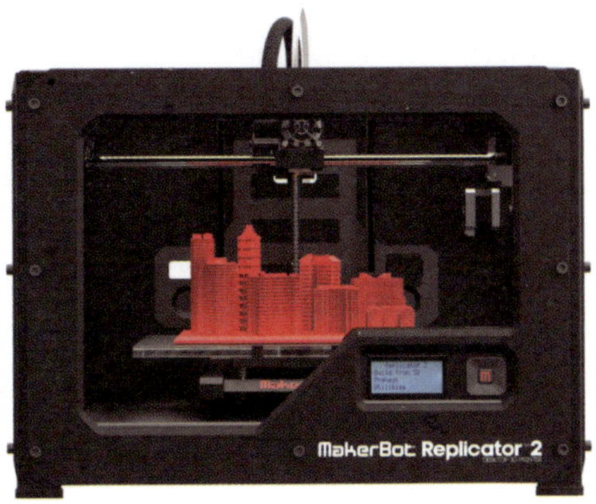

Design & Features

Departing from the hobbyist-kit approach of earlier 3D printers, the Replicator 2 has a clean, appliance-like aesthetic that feels at home on a desk or in a design studio. Its form signals approachability and professional performance. The powder-coated steel chassis is a rigid, industrial-grade structure that enhances stability, minimises vibration, and enables the fast, accurate performance that defines a new benchmark in desktop fabrication.

Inside, it features a dependable, high-precision single-nozzle print extruder engineered for efficient FDM (Fused Deposition Modelling). Working layer by layer with biodegradable PLA (Polylactic Acid) filament, the system delivers consistent and detailed 3D prints. A built-in LCD screen and five-button control pad support untethered operation, eliminating the need for a constant computer connection.

Typical Workflow

1. Begin with a 3D model, saved as an STL or OBJ format file, created in CAD software or sourced from an online community or store.
2. Import the model into MakerBot Desktop software, *slice* the geometry into thin sequential layers and generate code to control the printer's motion. Upload to the printer in a MakerBot-specific format (in 2012, standards were still being developed).
3. Load the spool (S) and thread filament (F) into the print head (H). Prepare the acrylic build plate (P) with a layer of blue painter's tape (this simple surface treatment improves adhesion during the first layers and helps with part removal after printing).
4. Start the print. After confirming that the first layer has adhered cleanly, leave the machine to build up the object layer by layer.

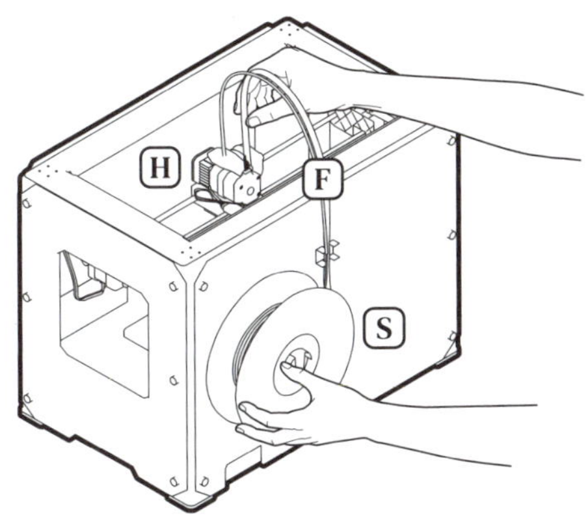

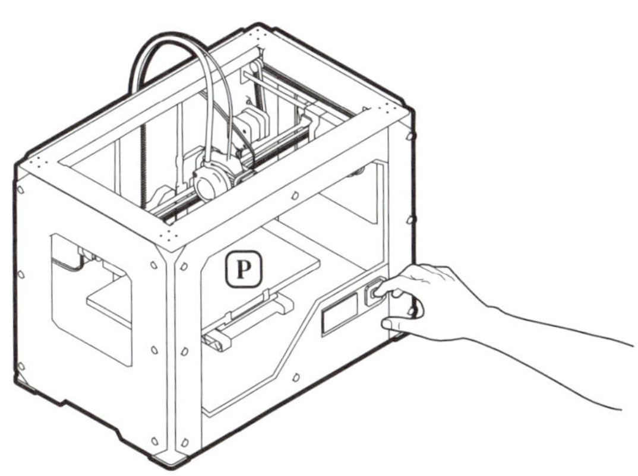

▲ *Cinematic Shot of Mountains in Scotland*

ARRI ALEXA Cultural Impact

In 2010, the film industry stood at a crossroads: digital capture was feasible, but many found that the resulting images had limited dynamic range, artificial or video-like colour, and lacked the natural grain structure familiar to film. The ALEXA's design prioritised these aesthetics. Its image sensor, colour science, and workflow reproduced the way photochemical film handles light, preserving shadow detail and delivering highlights that roll off smoothly rather than clipping suddenly. Face details were retained, and skin appeared softer and more dimensional.

Cinematographer Sir Roger Deakins ASC, a longtime champion of traditional film, shared his experiences. After testing the ALEXA, he felt the images had a quality comparable to film but a dynamic range superior to that of film. His decision to use one to film James Bond: Skyfall (2012) served as an endorsement of what was becoming possible in the digital era, as evidenced by the overcast, panoramic mountain scenes filmed in Scotland. Deakins appreciated how the camera captured more latitude and an enormous amount of detail in the shadows. He also valued the confidence that came from connecting the camera's digital output to accurate on-set monitors, which enabled precise discussions about lighting and silhouettes that had previously required overnight film processing.

Cinematographers could now rely on digital technology to push the limits of lighting and visual ideas, seeing precisely what the camera would capture. Directors could review takes immediately on set, making more dynamic and efficient decisions. Multiple Academy Awards for Best Cinematography and Best Picture have been awarded to films shot on ALEXA camera systems, and nearly all film production has now shifted to digital cameras.

Teenage Engineering OP-1 Cultural Impact

The OP-1 brought a sense of childlike wonder and playfulness in a period when synthesisers had become technical and complex. Instead, it offered curiosity, immediacy, and surprise, reframing how a creative tool could be packaged and valued.

Co-founder Jesper Kouthoofd described the company culture as akin to the energy of being a teenager[12]: an eagerness to experiment and a mindset of simply trying, rather than worrying about a lack of knowledge or not having done it before. Co-founder David Möllerstedt explained that the OP-1 was not created to conform to existing market expectations. Instead, it sought to redefine how a synthesiser might behave, capturing the magic of performing and recording in a linear and physical way. They deliberately chose to design a digital tool that limits functionality, challenging the prevailing belief that more functions equate to greater creative freedom. In short, the OP-1 carefully used constraints to reduce potential distractions and encourage focus, thereby stimulating spontaneous creativity. Möllerstedt also emphasised the importance of designing tools that feel good to own and use, as well as sound good. The priority was to create a joyful user experience that reconnects people with creativity itself, even at the expense of traditional specifications and features.

Many users regard the OP-1 not only as a tool but also as a source of fascination and inspiration; some simply enjoy turning it on and seeing what unfolds! It fostered a culture that values the presence, curiosity, and playfulness of the artefact itself, alongside the creative expression and output it enables, and encouraged people not only to make music but also to rediscover the pleasure of making anything at all.

MakerBot Replicator 2 Cultural Impact

The Replicator 2 launched amid a controversial change in the business model for creative tools serving the maker movement. Before its release, MakerBot championed the RepRap (Replicating Rapid Prototyper) philosophy, open to build-it-yourself cloning. With the Replicator 2, they would sell fully assembled units, using a commercial model that retained proprietary control over design, manufacturing, and software. At first, the decision created a deep cultural rift. Community veterans felt betrayed, arguing that MakerBot had built its success on the back of shared public knowledge. However, it also stabilised the marketplace. A preassembled, reliable tool that worked out of the box attracted a new wave of users who cared more about the final products they could produce than the mechanics of the machine. Bre Pettis, then-CEO, said the Replicator 2 was the first affordable 3D printer that did not require specialist knowledge to operate,[13] moving users from tinkering to making, with ease of access central to its impact.

Online design-sharing sites such as Thingiverse emerged, enabling users to share thousands of printable models. MakerBot cofounder and Thingiverse creator Zach Smith saw this as a necessary platform to accelerate innovation, remove distribution barriers, and democratise access to physical design files. A wave of localisation began, designers able to reliably prototype products in their own offices, protecting their intellectual property and eliminating the gap caused by shipping physical samples from factories.

From low-cost Robohand prosthetics to the challenging possibility of 3D-printed weapons, the power of manufacturing was no longer confined to factories. It was now prepackaged as easy-to-use tools that ran on desktops.

50 | OpenAI DALL•E

Category: Illustration, Design, Photography

Release: 2021

Designer: OpenAI

Manufacturer: OpenAI, North America

Dimensions: N/A
(billions of mathematical model parameters)

DALL•E marks a turning point in creative tools, introducing machine intelligence that transforms words into images at scale. Developed by OpenAI, their generative model converts text prompts into digital image compositions, ranging from realistic photographs to stylised illustrations. Rather than reproducing stored image pixels, DALL•E works in an abstract conceptual space, recombining learned ideas and relationships. It can interpret sentence logic, infer context, connect concepts, and generate imagery that is both meaningful and visually coherent.

As important as the underlying model is, the way DALL•E functions as an interactive system is equally important. By using language, the interface is democratised, enabling anyone to easily act on their ideas.

Design & Features

Trained on millions of text-image pairs, DALL•E learns how words, objects, and stylistic cues correspond to visual concepts. A mathematical model organises this knowledge into an abstract and compressed *latent space*, enabling the generation of unique compositions by blending distinct ideas into visually coherent hybrids. Users provide written prompts, where careful attention to phrasing can guide content, composition, style, and emotion. Re-prompting can refine and expand, allowing users to add, remove, or extend elements within an image.

DALL•E fueled ongoing discussion about authenticity, ownership, safety, and bias. Ethical design, including watermarking, prompt filtering, bias adjustments, and restrictions on depicting public figures, attempts to address some of the complex legal, moral and social implications of data sources, image generation and end use.

Typical Workflow

1. Articulate the concept, style, and emotional direction for your image using a written prompt. This could be literal, such as 'a modern kitchen interior', poetic, like 'a dream of light and shadow', or playfully imaginative, like 'a cute wasp eating a moist blueberry'.
2. DALL•E interprets the prompt and generates multiple visual possibilities.
3. Adjust the wording to refine details: maybe alter the background, or shift the lighting?
4. Close the creative loop with curation: select the most compelling output and re-prompt to explore variations and compositions. Download the final image. Additional adaptation can be made using traditional photography and design software to extend and finalise the work. Decide if it is important to tag your work as *AI-generated*.

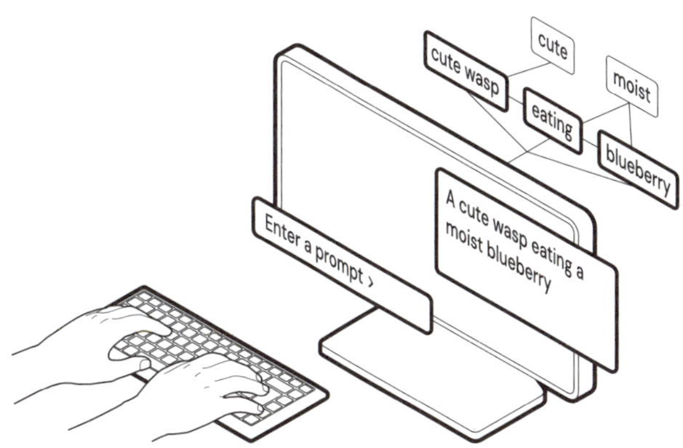

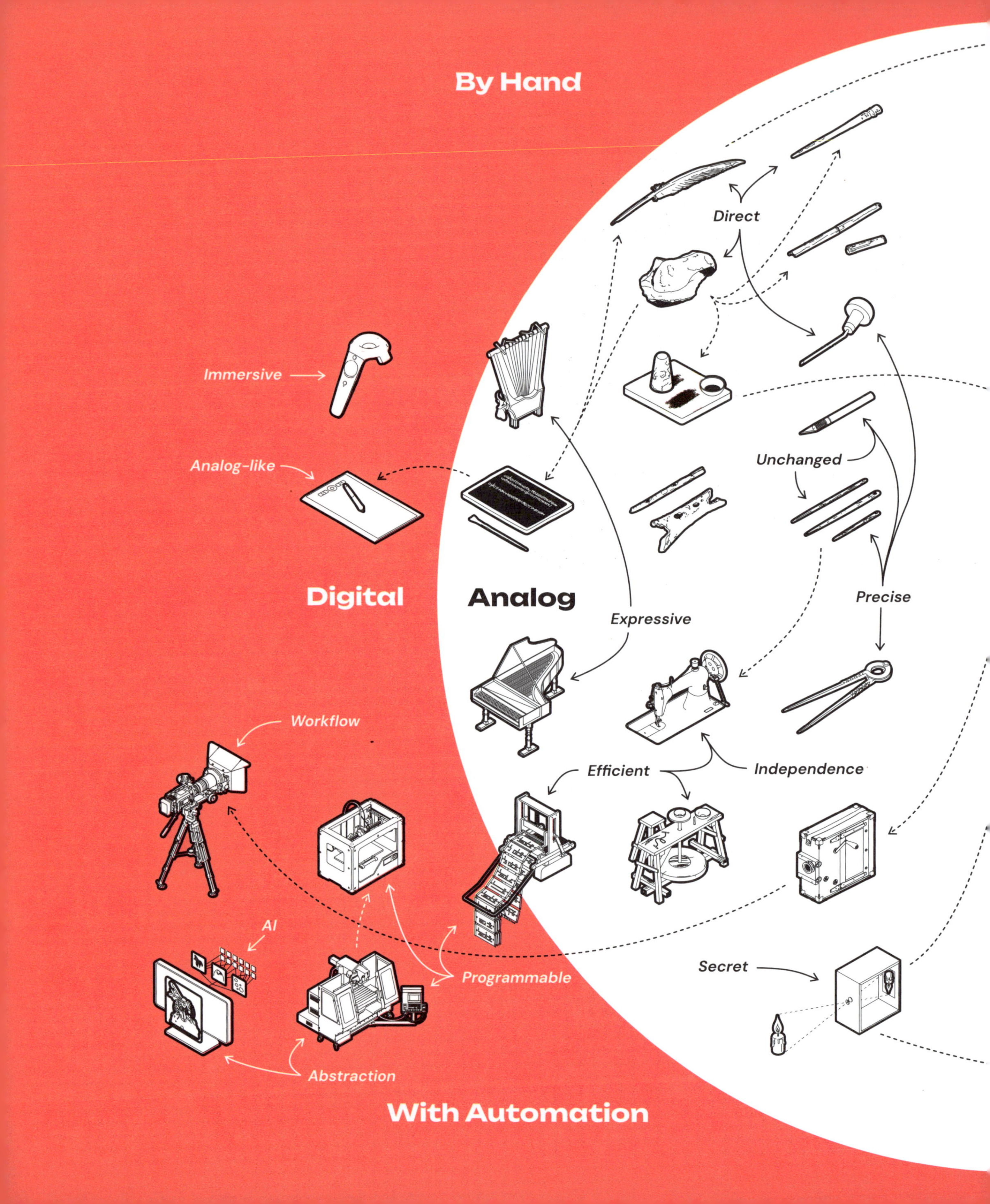

By Hand
Direct
Immersive
Analog-like
Unchanged
Digital
Analog
Expressive
Precise
Workflow
Efficient
Independence
AI
Secret
Programmable
Abstraction
With Automation

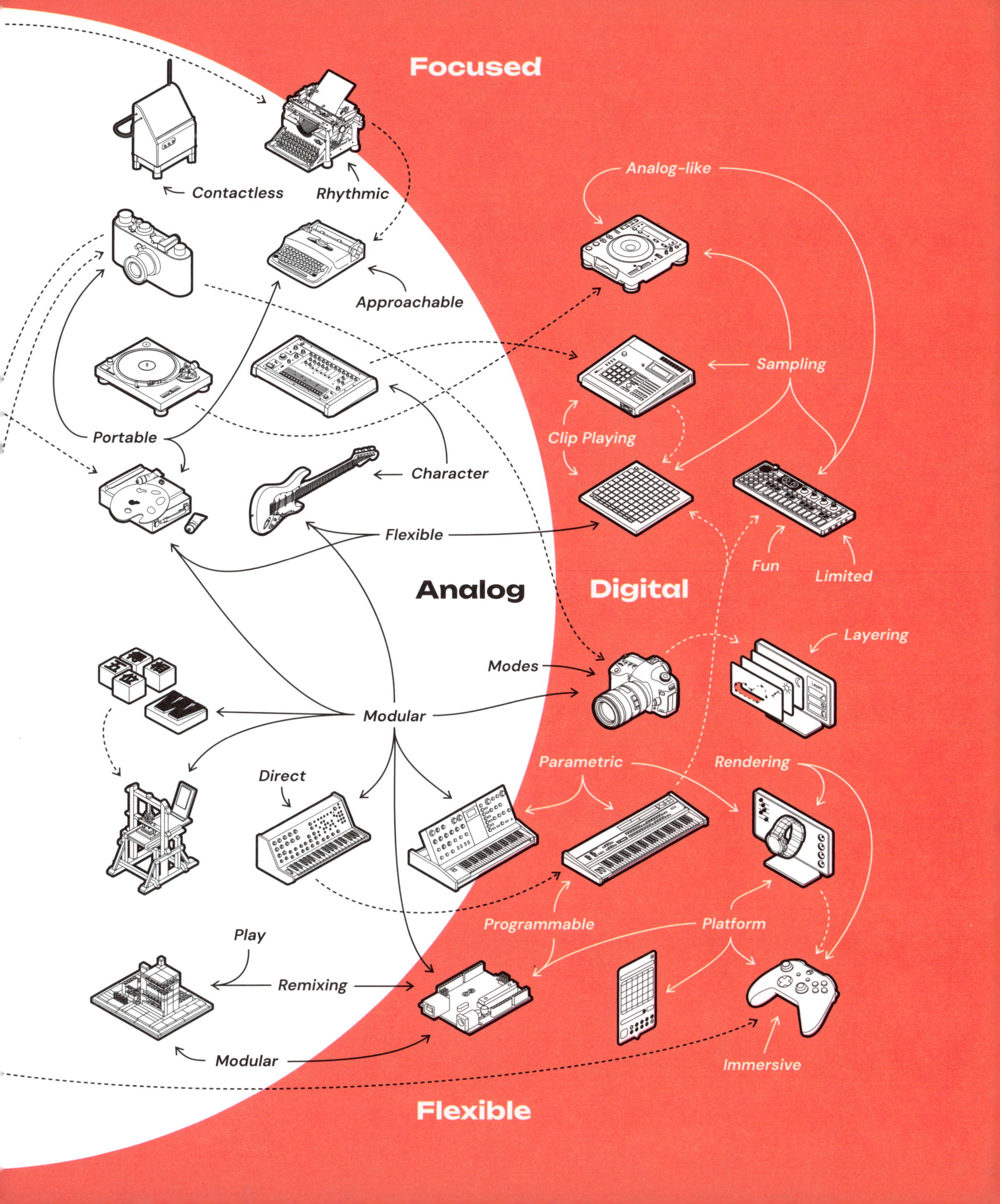

Focused
Contactless
Rhythmic
Approachable
Portable
Character
Flexible
Analog
Digital
Analog-like
Sampling
Clip Playing
Fun
Limited
Layering
Modes
Modular
Direct
Parametric
Rendering
Programmable
Platform
Play
Remixing
Modular
Immersive
Flexible

Open AI DALL•E Cultural Impact

DALL•E redefined who could participate in visual creation. Before its release, producing high-quality imagery required technical training, specialised tools, and, for photography, a real location and subject. With DALL•E, language became the primary interface, meaning the ability to articulate an idea mattered as much as drawing skill or knowing how to stage and capture a good photograph. Educators, knowledge workers, and business owners began generating teaching materials, illustrations, and marketing images without commissioning an expert designer. Journalists experimented, creating editorial imagery that revealed just how quickly these new synthetic visuals could become surreal, persuasive, and outright misleading.

The tool's design shaped its impact. DALL•E rewarded users fluent in art history, composition, and visual references. Its early limitations introduced an aesthetic of their own, with awkward anatomy, broken text, and odd physics. The upside was speed and access. The downside followed just as fast. Critical questions regarding authorship, originality, labour displacement, and the environmental cost emerged as use increased. Visual abundance replaced visual scarcity, and media platforms became saturated with disposable imagery optimised for engagement and, worse, misdirection and lies. AI-generated *slop* had arrived.

Ownership, attribution, and fair compensation at scale remain significant cultural and technical challenges, as do significant resource use and environmental impact. Any redefinition of the role of creative labour should be challenged, to highlight the importance of preserving human-led, experience-based, handmade work as a cultural signal and of designing generative tools that add meaningful and lasting value.

[1] Amelia Gentleman, 'A City That Sat on Its Treasures, but Didn't See Them', *New York Times*, March 19, 2008, nytimes.com/2008/03/19/world/asia/19chandigarh.html.

[2] Walter J. Ong, Orality and Literacy: *The Technologizing of the Word* (1982; repr., London: Routledge, 2012), 77–79.

[3] Wolfgang Amadeus Mozart to Leopold Mozart, October 17, 1777, in *The Letters of Mozart and His Family*, ed. Emily Anderson, 2nd ed. (London: Macmillan, 1966), 1:329.

[4] Pierre-Auguste Renoir, quoted in Jean Renoir, *Renoir, My Father*, trans. Randolph and Dorothy Weaver (1958; repr., New York: New York Review Books, 2001), 69.

[5] Mary Brooks Picken, *The Secrets of Distinctive Dress: Harmonious, Becoming, and Appropriate Dress* (Scranton, PA: Woman's Institute of Domestic Arts and Sciences, 1918), 7.

[6] Clara Rockmore, *Method for Theremin*, ed. David Miller and J. S. G. Miller (1987; repr., San Diego: Wrightwood Press, 1998), 12.

[7] Eric Clapton, interview by Tony Bacon, in *The Stratocaster Chronicles: Celebrating 50 Years of the Fender Strat*, ed. Tom Wheeler (Milwaukee: Hal Leonard, 2004), 114.

[8] Wendy Carlos, *Switched-On Bach*, Columbia Masterworks MS 7194, 1968, LP.

[9] Air, *Moon Safari*, Source/Virgin 7243 8 44978 2 8, 1998, compact disc.

[10] Afrika Bambaataa and the Soulsonic Force, 'Planet Rock', Tommy Boy TB 823, 1982, 12-inch vinyl single.

[11] Massimo Banzi, 'How Arduino Is Open-Sourcing Imagination', *TEDGlobal 2012*, filmed June 2012, video, 15:45, ted.com/talks/massimo_banzi_how_arduino_is_open_sourcing_imagination.

[12] Jesper Kouthoofd and David Möllerstedt, interview by David Z. Morris, 'Teenage Engineering's OP-1: The Most Important Synth of the Decade', *The Atlantic*, October 14, 2011.

[13] Marcus Fairs, '3D Printing Is "Bringing the Factory Back to the Individual"', *Dezeen*, 19 October 2012, dezeen.com/2012/10/19/3d-printing-is-bringing-the-factory-back-to-the-individual/.

4 | The Future

Understanding where creative tools are headed requires examining the most important trends and how they interact. New popular forms of creation have brought a fundamental change in what we perceive as original creation. This change is largely fuelled by the automation of tools, which seeks to satisfy a new generation of creators presumed to prioritise efficiency over originality, and artificial intelligence, which acts as the great accelerator.

Trends have a particular effect on human psychology: sooner or later, we are tempted to extrapolate them, usually through a technological lens, which leads us to embrace a deterministic vision of human evolution. However, it is less the way in which we look into the future that should concern us, but rather the fact that, by trying to predict it, we are already assuming the future will take a certain form — irrespective of what we want and no matter what we do.

The modern meaning of design builds on the idea of *intentionality*. Even when external factors are considered, the future is largely the result of the decisions we make today. Hence, how we design creative tools is never neutral and always promotes a certain idea of how we want creation, and thus culture, to evolve. In the 2011 film *Mr Nobody*, Nemo, the hero, creates an alternative reality when he makes important decisions, leading to multiple dimensions of his life coexisting in parallel.[*1] When we embrace the status quo, we add a brushstroke to a certain vision of the future, which, at present, increasingly resembles a situation in which creation is entirely automated, commodified, and regimented to fit specific *consumer* needs. Conversely, every time we question established archetypes, explore ideas for doing things differently, or pursue craft and creative processes that embrace happy accidents, we open up new possible futures. And who knows, one of these might turn out to be our *actual* future.

---

[*] While the narrative presents these lives as parallel realities, the film eventually reveals them to be the mental simulations of the protagonist as a child, imagining his future.

From Collage and Sampling to AI

'Collage' — literally 'gluing' in French — is believed to have been coined around 1912 by Cubism founders Georges Braque and Pablo Picasso to describe visual artworks made from paper, fabric, and other materials.[*2] The technique, however, likely originated in the 11th century in Asia, and has been practised in Europe since at least the 18th century, notably by British artists Mary Delany and Mary Georgina Filmer.[34] In 1973, American musician DJ Kool Herc started reusing musical phrases — or *samples* — from vinyl records, manually looping them on turntables in order to form the basis of new tracks, in what many consider as one of the founding events of hip-hop.[5] Other artists followed, and *sampling* became a fixture of the genre.

While apparently unrelated, collage and sampling[†] proposed a new idea of creation in which creators could take bits here and there, chop, rearrange, and join them together in order to make new art — instead of painstakingly learning how to imitate nature or the work of others, as had been the case before. Today, these techniques form the foundations of a new creative paradigm that prioritises point of view and, at times, recontextualisation, over the mastery of foundational skills such as drawing, painting, or playing an instrument. Its ramifications are felt across most aspects of modern culture, from music remixes to internet memes and, to some extent, the growing popularity of photography.[‡]

There is a striking parallel between, on one hand, collage or sampling and, on the other, artificial intelligence. Indeed, whether we call it art or just an *output*, what comes out of generative AI is a mashup of what it has been fed; only, the pieces seem to have become so small that we can no longer recognise them[§] — which makes questions about authorship and ethics all the more complex. This parallel has been noted by many, including Canadian author Stephen Marche, who sees it as a sign that AI may facilitate more *derivative art* and thus 'be a thread to lead us out of the labyrinth of the formulaic'[6] — as sampling did for music when it changed the course of hip-hop.

But what if the probabilities for AI to maintain us in the 'labyrinth of the formulaic'[7] far outweigh the probabilities it leads us out of it? Predicting the positive effects of AI, based on how sampling changed the course of music history in 1973, makes for a nice story. Yet it also omits key details: at the time, sampling required considerable work: first, finding interesting instrumental parts, and then looping them using turntables, after taking the time to master the not inconsiderable skills required. If drum machines and samplers made the process easier in the 1980s, they did not make it *easy*, and certainly not quick.

---

[*] In reality, they likely used 'papier collé' — French for 'glued paper' — at that time.

[†] And other derivative methods of artistic production, often with names taken from French — notably 'assemblage', the creation of three-dimensional works from found objects, and 'bricolage', which involves constructing art from a diverse range of available materials.

[‡] Photography served as an early catalyst for this paradigm shift, sparking intense 19th-century debates over whether mechanical reproduction could constitute 'art'. However, because it remained focused on the capture of a singular, 'whole' image, it was not until the advent of collage and sampling that the 'bits and pieces' approach — rearranging existing fragments into a new whole — emerged as a dominant creative logic.

[§] Metaphorically a mash-up, though technically a projection: generative models do not contain or rearrange original data. They are predictive engines that output content by calculating the mathematical likelihood of patterns observed during training.

'what enabled creative divergence was the use of creative tools that left room for a form of "human contamination".'

Hence, what, for decades, enabled collage and sampling to be used truly creatively was the unpredictability of human creators, which would shine through a meandering process: how one individual might associate various elements to form a visual or musical collage was unique to how they experienced, remembered, and thought about the world. In other words, what enabled creative *divergence* was the use of creative tools that left room for a form of 'human contamination'.

However, creative tools changed, and they did so long before the introduction of generative AI. In an article for *Common Edge*, Natalia Ilyin[8] recounted her experience as a judge for design competitions organised by the American Institute of Graphic Arts, comparing the mid-1990s, when graphic designers were still using pencils, knives, and masking film, to the present day. At some point, when software tools had become the only tools used by the designers submitting their work, she noticed: 'Most of the work we looked at was similar in textures, colours, and typography. Some of it was very good . . . some of it was very well thought out. But most of it seemed as though it had been made by the same person. By the same consciousness.'[9]

While she views it not merely as a technical shift,[*] Ilyin attributes this uniformity to the designed constraints of software applications, which force designers to 'comply with a world of snap-to grids' and 'prescripted ways'. One might also add the massive acceleration of the creative process and the introduction of 'copy-paste', both of which carry a particular resonance here. With the 'softwarisation' of tools, collage, sampling, and derived techniques have become much simpler and more popular means of creating with reduced effort. Today, most creative tools, especially in music and the visual arts, encourage the use of premade elements, including templates, music samples, stock images, filter packs, and 3D assets.

For aspiring artists, the temptation is strong to build from these rather than risk the frustration of unmastered techniques and the disappointment of amateurish results. Yet, each shortcut is also a missed opportunity to develop a craft and, with it, a unique style or voice — which maintains them in a state of dependency. On a larger scale, it reinforces stereotypes of how things 'should' be, instead of encouraging diversity in how creativity is expressed. This is one reason many artists create work that looks or sounds the same: when trying to imitate what they like, they can easily do so, sometimes by simply reusing the same source material. As a consequence, what used to be a nod, a pastiche, or simply fan art is becoming indistinguishable from what inspired it.

Hence, there is no reason to think that AI — on its own — will bring back the creative effervescence of hip-hop in 1973, or Cubism in 1912, and lead to another meaningful revolution in *derivative art*. If anything, it will simply offer more technical solutions for copying and reuse. Ironically, the ease with which AI enables everyone to sample and collage the world is also what threatens to eliminate the opportunities for accidental 'human contaminations' necessary for creative divergence and true *derivative art*. In the long term, this poses a real risk of a global inward collapse of creativity.

* Ilyin develops the idea that it is a 'system of indoctrination' that prioritises corporate systems over individual genius, and she notes the sociological evolution of the creator driven by the high cost of these increasingly expensive computer tools.

But AI is a technology — *not a tool.*[*] In theory, nothing prevents the development of creative tools that use AI in different ways and favour long journeys over formulaic results. Similarly, sampling and collage are just ideas: they don't have to be the basis for the commodification of creation. In other words, one must look beyond current technologies and ideas — in effect examining how tools are *designed* — to understand where creativity is headed, for these are merely symptoms of a broader phenomenon: a subtle, yet profound, shift in the very meaning of the word *creation*.

Culture Extrapolation

'As culture becomes completely commodified, it becomes the star commodity of the Spectacular Society.'

— Guy Debord, 1967[†10]

We should pay close attention to the content creator phenomenon, for in sheer 'volume of participants' it has already surpassed every other trend or creative movement ever to exist.[‡] This has been made possible, in large part, by the commodification of creation, which enables a new generation of creators — *content creators* — to earn a living from their work, primarily through advertising. It also represents a major opportunity for larger stakeholders: from the social media platforms where content is consumed to AI technology providers, and creative tool manufacturers — who all seem too preoccupied with capitalising on the situation to ask questions, even if the threat looming on the horizon is a flattening of creativity and, further down the line, culture as a whole.

At the turn of the 2020s, many companies began targeting their communications toward content creators, and not only the social media platforms, which require content to drive what they call *engagement*. Today, *pro* smartphones dominate manufacturers' marketing campaigns with annual technological reveals that entice with beautifully staged CGI breakdowns and a bombardment of jargon for features that, we are told, were once accessible only to *professionals*. Despite a typically much higher price point, *pro* smartphones often outsell standard models[11] — and not just because humans like to impress others.

In a 2023 study, Goldman Sachs Research estimated that the *creator economy* was worth about $250 billion and would double by 2027.[12] Illustrating this trend, an often-cited study by LEGO stated in 2019 that 'children were three times more likely to aspire to be a YouTuber (29%) than an Astronaut (11%).'[13] For creation at large, the revolution happening before our eyes may have worse consequences than we think.

On the surface, we seem to be transitioning from a hierarchical, centralised system, in which a few — typically large film production studios or established musicians — created for the masses, to a distributed model in which more people create and generate revenue. In theory, this should lead to more diverse output of human creativity. In practice, what is 'worthy of being created' is increasingly regimented by a ruthless model in which content creators need

---

[*] As developed in the introduction, while often used interchangeably, technology and tools represent distinct concepts. Technology is the practical application of scientific knowledge; a tool is a specialised object designed to facilitate a specific task. Thus, while technology defines the scope of what is possible, a tool is the result of design decisions regarding how to apply that potential.

[†] By 'Spectacular Society' Guy Debord meant a society merely obsessed with images or media.

[‡] Numerous sources report hundreds of millions of content creators. Never in human history have so many individuals been simultaneously engaged in a self-identified creative pursuit. The 'Maker Movement' comes to mind, but while that was defined by physical hobbyism, the content creator phenomenon aggregates all media — including an increasing number of self-declared 'makers'.

to imitate the few successful individuals among them to get a chance to be seen and survive, adapting their creation to the advertising-driven platform metrics they are fed in real-time.*[14] As writer Taylor Crumpton puts it in *Time* magazine: 'Social media platforms reward visibility. Algorithms don't distinguish between authenticity and performance — they simply amplify what's most clickable.'[15]

Surely, there are still people who paint, write, make furniture, and compose music, not for sharing on social media but for themselves, their friends, or the sake of the craft, preserving a living culture where the creative act matters more than the recognition. After all, content creation does not represent *all* creation. Yet the fact that it is taking up increasing amounts of space is hard to ignore. Who, today, doesn't naturally associate the term 'creator' with 'content creator'? Aren't we risking the slow disappearance of the *traditional* creator when the world's attention — and capital — is increasingly siphoned into the content economy?

By pretending to pursue the goals of making people more creative and 'democratising creation', companies that benefit from this shift are, in effect, restricting the very meaning of creation and turning it into one of the most prosaic forms of human activity. And those who dismiss this shift and hide behind classificatory debates on what constitutes *true* art miss how it has already affected culture at large. In 2025, an article in *The Guardian* by British writer and critic Phil Hoad outlined the deep implications of the rise of 'algorithm movies' — films produced for streaming platforms where sound mixes are flattened, photography homogenised toward low-contrast tones, and plots simplified, with characters literally announcing their every move to keep the attention of an increasingly distracted audience.[16]

Similarly, in *Filterworld*, American writer Kyle Chayka refers to the work of scholar Kate Eichhorn to describe a new paradigm where aspiring artists must build 'content capital'[17] to attract an audience *before* they even begin their actual work: 'the emphasis is not on the thing itself but the aura that surrounds it, the ancillary material that one produces because of living the lifestyle of a creator. That ancillary content might be Instagram selfies, photos of a painting studio, evidence of travel, tossed-off observations on Twitter, or a monologue on TikTok. It all builds an audience for the person, who remains a separate entity from the things that they make.'[18]

The danger, then, is the erosion of the environment that allows creators to exist in their full diversity. When attention becomes increasingly scarce and is redirected toward the shallow, the spectacular, and the entertaining, then the great flattening of creation threatens culture in its entirety. Now that the *creator economy* has given birth to an established market, its main economic actors need it to grow. In this context, the automation of creation appears to be a significant market opportunity.

Creators, aspiring and accomplished, are sold a tale: inside each of them resides a genius visionary whose sole task should be to focus on having ideas, for machines can take care of the supposedly lesser important act of making. This clever narrative plays into the 'hidden genius' trope — the comforting myth that the only thing keeping someone from revealing their great artist potential is a set of skills that have been rendered unnecessary by technology. But behind this illusion lies a less glamorous possibility: when creation is entirely commodified and art reduced to content, human creators will no longer be the driving force, simply a *bottleneck*.

* According to the Goldman Sachs study cited above, only about 4% of global creators earn more than $100,000 per year.

It won't be necessary to 'unleash the creative potential' of anyone, since automated creation will be able to feed humanity's bottomless appetite for entertainment in a much more effective manner. In this scenario, individuals will be fed what they desire — music, films, and video games — all entirely generated and adapted to their taste using one form or another of AI. Ultimately, culture will cease to be the result of what we collectively create: good and bad, limited by our ideas and skills, and shaped by the unpredictable, the misunderstandings, and the happy accidents. Instead, it will become an extrapolation, a generative prediction — in short, a best guess of what it could have been.

The End of Creative Tools?

'I live on Earth at present, and I don't know what I am. I know that I am not a category. I am not a thing — a noun. I seem to be a verb, an evolutionary process'

— R. Buckminster Fuller, 1970[19]

Projections based on past trends lead us to believe that we can predict the future. But the very idea of design posits precisely the opposite: our future is the choices we make today.

On one hand, it seems that the companies behind the most widely used creative tools have entangled themselves with a market dynamic that incentivises them to continuously push for more automation in order to grow their user base, in a race against 'AI-first' competitors.[*20] From this perspective, both older and newer actors might decide they are no longer in the business of *creativity* after all, but in the business of *creation* itself. Such a repositioning would likely signal a transition toward a model in which the only humans worth considering are those who consume what is automatically generated for them by AI.

'when creation is entirely commodified and art reduced to content, human creators will no longer be the driving force, simply a bottleneck.'

On the other hand, if these companies stop making creative *tools*, it would not necessarily mean the end of such tools. Today, many small companies and individuals work on their own ideas. Moreover, if some of the most iconic tools have historically come from independent tinkerers and entrepreneurs, development and manufacturing means have become, on the whole, more accessible than ever before. Ironically, this is in part due to the transition to digital, software, and AI technology — which shows that a purely technological lens can never really tell us where the future is headed.

Therefore, the risk is not so much the lack of agency afforded to the designer but rather the dominance of simplified narratives and preconceived ideas that constrain imagination. Thinking that a creative tool is *necessarily* a software tool, or that certain technologies call for certain interfaces, is the reason we seem to get stuck thinking that the future of creative tools *necessarily* implies a chatbot — or, if we're lucky, a node interface. Designing new creative tools is not necessarily going in one direction, and certainly not only toward simplifying everything.

When Bartolomeo Cristofori created the first piano, the design was largely inspired by what

[*] In a 2023 interview for Stratechery, Adobe then Chief Strategy Officer Scott Belsky provided rare insights on how this situation came to be, explaining that Adobe built its AI strategy both to counteract 'AI-first' competition and to capture a broader user base of 'non-professionals' who are presumed to rely on automation rather than manual skills.

had been in use until then: the harpsichord. Yet one of his crucial contributions was to give the musician the ability to strike each key with varying force — hence the early name *fortepiano*.* This admittedly made playing the piano considerably more difficult and potentially slower than playing the harpsichord. In fact, as the legend has it, Johann Sebastian Bach was rather unimpressed the first time he tried, finding the instrument 'too hard to play.'[21] Yet the design also resulted in a broader palette of possible nuances and a richer sound. Today, the piano is still a popular instrument, whereas the harpsichord has been relegated to museums and period drama.

There are multiple ways for a designer to invent the future of creation without necessarily falling into the prosaic or starting from a blank canvas. Designing a creative tool demands practical questions and natural curiosity — typically cultivated by looking at and learning from a wide range of devices and interfaces, past and present, that provide an inspirational starting point and useful lessons upon which to explore future ideas.

Future Interfaces

If we consider software tools, they typically rely on a laptop, smartphone, or tablet as their *primary* interface, which, in turn, determines their modalities of expression and the types or styles of work creators perform with them. Typically, tools like Illustrator have evolved to take advantage of mouse and keyboard interfaces — offering precision, speed, controllable curves, and fundamental shapes — while tools like the app Procreate embrace touch and a stylus, offering greater opportunities for improvisation and immediate expressiveness.

Despite constraints, a growing proportion of new creative tools are software-driven, as they have lower start-up costs and are easier to distribute[22] — which is logical since most people own a laptop, smartphone, or tablet. It therefore raises an interesting question for the future of software tools: What if the common devices were replaced by new devices with different interfaces? What if our smartphones were replaced by smart pendants or pairs of AI-powered earphones? What if our laptops and tablets were replaced by AR† glasses, headsets, and brain-computer interfaces?‡

There are at least three routes by which these questions can be explored. The first, obvious question is how it would affect the modalities of expression. A pendant, or a pair of earphones, could run the best AI model in the world, but if the only primary interface is a microphone, any creative tool running on them will have to rely on speaking, singing, humming, or maybe finger-tapping. Minimalist AR glasses and immersive headsets offer the possibility to work spatially, which may enable more natural interactions. As for brain interfaces, they may enable creation to be transformed into a form of active dreaming.

Of course, designers don't have to accept the modalities of expression imposed by any new

* Originally gravicembalo col piano e forte — meaning 'soft and loud' — the name was shortened to 'fortepiano', 'pianoforte', and, by the mid-19th century, simply 'piano'. As the instrument became standardised, 'fortepiano' emerged as the term for the Classical-era artefact, distinguishing it from the modern, cast-iron iteration.

† Augmented Reality: a technology that superimposes computer-generated images, sounds, or other data onto a user's view of the real world, providing a composite view. This differs from Virtual Reality (VR), which creates a completely immersive synthetic environment, and Mixed Reality (MR) — a hybrid of the two — which allows digital and physical objects to coexist and interact in real time.

‡ A brain interface, or Brain-Computer Interface (BCI), is a system that creates a direct communication pathway between the brain's electrical activity and an external device, translating thoughts or neural signals into commands to control computers.

▲ ► *Designed for use with headsets or smart glasses, the InSight camera eliminates the need for a screen or markings, as it does not require direct observation. The viewfinder, frame overlay, and settings are augmented directly in front of the photographer, allowing maximum freedom in how the camera is held, thus encouraging creative exploration in framing and composition.*

device that ends up becoming a dominant one. By *designing* a dedicated creative interface for their tools, they can make use of the large palette of what Howard Gardner calls our *multiple intelligences*.[23] In 2013, FiftyThree developed their own stylus — named 'Pencil' — to work with their illustration application for iPad, preceding the release of Apple's own Pencil by two years. In music-making, most manufacturers of software tools propose their own physical interfaces, such as *Maschine* by Native Instruments and *Push* by Ableton, that provide a more kinaesthetic experience of creating music. Who knows, the tools of tomorrow could take the form of dedicated interfaces that connect seamlessly to software running on a tiny pendant or a pair of earphones!

Finally, some designers might find the dependence on a primary interface too constraining and use this as an excuse to build their own dedicated creative tools. Today, most cameras are effectively smartphones. This results in limitations that restrict their very design as creative tools: few dedicated buttons, questionable ergonomic choices, and small image sensors that do not perform as well as larger ones when it comes to essential artistic controls such as depth of field. If smartphones were to be replaced by smart glasses, the everyday* camera would either disappear or, more likely, be reduced to a functional tool that captures only what we see, from the direction we are looking — and likely controlled only by voice or gestures. That would be a tragic over-simplification of a popular practice led by a misunderstanding of the craft of photography.

Yet, like the smartphone before them, smart glasses offer an opportunity to elevate photography by making it easier for the photographer to frame and compose *spatially*. Hence, this transition could serve as a starting point for a complete reinvention of the camera archetype, in which the camera is redesigned to facilitate creative control and encourage exploration without requiring the photographer to look at — or through — the camera at all.

If these three routes are all valid, they also assume that today's most popular forms of

* As opposed to the larger cameras designed for professionals or serious hobbyists.

creation will remain unchanged. Yet, as photography, cinema, and video games have shown, there are constantly new ways in which humans wish to express their creativity. Sometimes, designing the tools of the future requires anticipating what will capture the imagination of the creators of tomorrow.

New Forms of Expression

In almost all fields, the base unit of creation is changing: whereas for centuries it was the brushstroke, the single word, or the single note, creation now encompasses larger pre-built modules, which enable creators to repurpose their creative energy and go further than ever before. Game designers, for instance, can put together a complex 3D world by combining assets designed by others with a few designed by themselves — typically the main characters or key props.

Sometimes, creative tools enable artists to transform what used to be a static work of art into an interactive piece. In a live performance today, musicians can manipulate their compositions in real-time, changing the sounds that they chose when they composed a piece, re-arranging entire parts of their work on the fly, and even inviting the energy of the room to dictate the direction of the music, ensuring that no two performances are ever identical.

This transition towards interactive media is far from having been fully explored, and potentially an interesting direction to take new technologies, not to replace human creativity, but to build new tools that can encourage and grow it. In filmmaking, the creative tools of tomorrow may enable directors to build truly interactive movies that respond to the audience by adapting the narration, just as a narrator can tell the same story 50 times and yet never twice exactly the same way. With the creative tools of today, shooting and editing such a movie would probably be too complex and certainly expensive to produce. With tomorrow's tools, small variations — such as alternative camera angles from the original footage or dynamic changes in focus — could be generated using AI, not instead of human creative work, but as an extension of it, opening up new creative opportunities.

This may also mean a more profound rethinking of how we design tools: a shift from rigid, manufactured instruments to ones that are essentially alive.

Living Tools

Most contemporary creative tools are 'static,' designed entirely by human hands. This carries a significant advantage: these tools are the distilled result of collective intelligence. When creators use a standard paintbrush or video editing software, they benefit from decades — sometimes centuries — of shared wisdom regarding ergonomics and workflow. However, these tools also impose specific ways of creating that can restrain intent, and many creators are hence driven to design their own bespoke tools.

The risk of such an approach is loss of the collective intelligence embedded in the idea of shared tools. We could, however, imagine an alternative model in which tools share a common platform of fundamental principles

▲ *Komanda was designed to create and showcase live, interactive film and animation. On stage, it acts as a conductor's baton for non-linear storytelling, enabling scene launching, camera control, and real-time compositing within the projected visuals. In 2017, it facilitated the world's first mixed reality concerts across London, Paris, and Amsterdam.*

that are then adapted to each individual or specific situation. In many ways, this is not very far from what platforms like Arduino or Lovable propose today — even if these also impose constraints of their own.

True 'living' tools would not require the artist to become a tool designer, for their properties could adapt dynamically. For instance, combined with the idea of sampling and collage, properties could be inferred from a sampled artefact which, instead of being a passive witness of what was, would carry — in some form of creative DNA — the entire toolset required to recreate or mutate it in new and unique ways. In the context of digital music, today's static tools treat a music sample as a frozen audio file on which only a limited amount of manipulation can be done. In a 'living' tool, that same sample could carry the 'DNA' of its origin: for instance, a recreation of all the instruments used in the recording, right down to the properties of the studio on that specific day. For 3D modelling, any real-world object could be used to infer the specific tools necessary to recreate it with as much detail as desired.

If that sounds impossible, there are in fact already several tools that can do a limited version of this.* Ultimately, in a world where people are too often encouraged to abandon creation in order to enjoy consumption, such

* Some tools can split audio into separate instruments, while others can reverse-engineer them entirely. For example, Synplant 2 uses an AI model called 'Genopatch' to revert any audio sample into synthesiser settings, allowing musicians to recreate, modify, and play that sound in a new composition.

tools would offer a constant invitation to take the DNA of the object of consumption and do just the opposite — create.

Designing the Future

Slogans have become the carbon offsetting mechanism of those who push for more automation and then pretend that 'what matters is the story you have to tell.' In *Enshittification*, Canadian-British writer Cory Doctorow describes the seemingly inevitable mechanism by which internet services become worse over time because of the platforms' need to capture more value than they create, driven by the pressure for unsustainable growth.[24] Similarly, the convergence of market interests and the automation of creative tools made possible by technological progress is taking us toward a point at which human creation might disappear entirely.

Luckily, unlike in Doctorow's *Enshittification* mechanism, creation does not have to obey market growth rules — and it does not need to turn into ready-to-consume entertainment. Hence, avoiding the commodification scenario will require reevaluating the 'content creation' narrative and our relationship with culture in general. However, to provide an alternative, we need creative tools to be *designed* — not with passive consumers in mind, but *truly*, for creators.

While *future interfaces*, *new forms of expression*, and *living tools* are all interesting directions, these are only examples of how designers can create a space to explore new ideas. For most of humanity's history, creative tools have come with friction, both literal and metaphorical, and limitations inherited from the technologies they relied on, that is, imposed by the laws of physics and the boundaries of human knowledge. It is the hallmark of progress that humanity eventually transcends the very constraints that once defined its reach.

'designers [must] ensure that their designs maintain an adequate distance between idea and result to allow for a genuine creative journey. From this one consideration, all the other design questions can then be derived.'

Tomorrow, AI might become the underlying technology in most human-made systems, further eroding the natural limitations of physics and opening a new paradigm in which few people can truly understand how these systems function. In the context of creative tools, this means a new key responsibility for designers: ensuring that their designs always maintain an adequate distance between idea and result to allow for a *genuine* creative journey. From this one consideration, all the other design questions can then be derived: Is it accessible? Does it allow for sufficient creative control? Does it support a wide palette of expression? Does it encourage the development of a craft?

If we do that, how humans express themselves might change, and the tools they use might evolve too. We might witness the rise of new, exciting forms of art and continue debating what constitutes true original creation, inspiring works, and respectable craft. There is no reason why human creativity should disappear, nor should our culture collapse with it. If we do that.

---

[1] *Mr. Nobody*, directed by Jaco Van Dormael (2009; Montreal: Les Films de l'Europe, 2011), cinema.

[2] 'Papier collé', Tate, accessed January 23, 2026, tate.org.uk/art/art-terms/p/papier-colle.

[3] 'Mary Georgiana Caroline, Lady Filmer', Art Institute of Chicago, accessed January 23, 2026, artic.edu/artists/106913/mary-georgiana-caroline-lady-filmer.

[4] 'Mary Delany', British Museum, accessed January 23, 2026, britishmuseum.org/about-us/british-museum-

story/people-behind-collection/mary-delany.

[5] Artur Galocha, '"I've heard that sound before!" How sampling influenced hip-hop', *Washington Post*, August 24, 2023, washingtonpost.com/entertainment/2023/08/24/ive-heard-that-sound-before-how-sampling-influenced-hip-hop/.

[6] Stephen Marche, 'The Future of Writing Is a Lot Like Hip Hop', *The Atlantic*, May 9, 2023, theatlantic.com/technology/archive/2023/05/generative-ai-novel-writing-experiment-stephen-marche/673997/.

[7] Marche, 'The Future of Writing.'

[8] Natalia Ilyin, '"AI Is Just a Tool"—and Other Fictions', *Common Edge*, February 18, 2025, commonedge.org/ai-is-just-a-tool-and-other-fictions/.

[9] Ilyin, '"AI Is Just a Tool"'.

[10] Guy Debord, *La Société du spectacle*, 3rd ed. (1967; repr.,Paris: Gallimard, 1992), thèse 193.

[11] James Rogerson, 'iPhone 15 Sales Figures Show the Pro Models Have Become the Default', *TechRadar*, May 2, 2024, techradar.com/phones/iphone/iphone-15-sales-figures-show-the-pro-models-have-become-the-default.

[12] 'The Creator Economy Could Approach Half-a-Trillion Dollars by 2027', *Goldman Sachs*, April 19, 2023, goldmansachs.com/insights/articles/the-creator-economy-could-approach-half-a-trillion-dollars-by-2027#.

[13] Kelly Nuckols, 'The LEGO Group Kicks Off Global Program to Inspire the Next Generation of Space Explorers as NASA Celebrates 50 Years of Moon Landing', *LEGO*, July 16, 2019,lego.com/en-us/aboutus/news/2019/july/lego-group-kicks-off-global-program-to-inspire-the-next-generation-of-space-explorers-as-nasa-celebrates-50-years-of-moon-landing.

[14] Goldman Sachs, 'The Creator Economy'.

[15] Taylor Crumpton, 'We're All Working for the Algorithm Now', *Time*, November 12, 2025, time.com/7332708/creator-economy-algorithm-unpaid-labor-privacy/.

[16] Phil Hoad, 'Bland, Easy to Follow, for Fans of Everything: What Has the Netflix Algorithm Done to Our Films?', *Guardian*, August 28, 2025, theguardian.com/media/2025/aug/28/bland-easy-to-follow-for-fans-of-everything-what-has-the-netflix-algorithm-done-to-our-films.

[17] Kate Eichhorn, *Content* (Cambridge, MA: MIT Press, 2022), quoted in Kyle Chayka, *Filterworld: How Algorithms Flattened Culture* (London: Heligo Books, 2024), chap. 4, 'The Influencer Economy'.

[18] Kyle Chayka, *Filterworld: How Algorithms Flattened Culture* (London: Heligo Books, 2024), chap. 4, 'The Influencer Economy'.

[19] R. Buckminster Fuller, Jerome Agel, and Quentin Fiore, *I Seem to Be a Verb* (New York: Bantam Books, 1970), 1.

[20] Scott Belsky, 'An Interview with Adobe Chief Strategy Officer Scott Belsky', interview by Ben Thompson, *Stratechery*, podcast, April 13, 2023, stratechery.com/2023/an-interview-with-adobe-chief-strategy-officer-scott-belsky/.

[21] Christoph Wolff, *Johann Sebastian Bach: The Learned Musician* (New York: W. W. Norton & Company, 2000), 413.

[22] Marc Andreessen, 'Why Software Is Eating the World', *Wall Street Journal*, August 20, 2011, a16z.com/why-software-is-eating-the-world/.

[23] Howard E. Gardner, *Multiple Intelligences: New Horizons in Theory and Practice* (New York: Basic Books, 2006).

[24] Cory Doctorow, *Enshittification: Why Everything Suddenly Got Worse and What to Do About It* (London: Verso Books, 2025).

Conclusion

Entire books have been written to analyse what motivates great artists; notable examples include Sigmund Freud's case study of Leonardo da Vinci.[1] Yet, while these discussions offer value in understanding the process, the *perspective*, and the influence of external factors, they can never explain why some are *woken at night* by the symphonies playing in their heads. Nor can they explain why others one day feel the urge to create entire worlds that they will spend the rest of their lives describing in the most intricate detail.

People create because they do — because they cannot help it.[*2] Designing new tools is not a rejection of old ones; it is a way of revisiting and expanding on them. In effect, once they reach a certain design maturity, the most successful tools evolve very little and remain with us, living through the hands of creators who one day fell in love. Will turntables survive for three hundred years like pianos have? Will cameras as we know them still be in use 30 years from now? Maybe. But if that happens because we have become incapable of proposing new creative tools that musicians, painters, writers, and all creators in general want to use — then it will be, first and foremost, a failure of design.

This book presents only a small selection of the hundreds of thousands of creative tools that humans have designed and brought to life throughout history. While considerable effort was devoted in making even the most unusual among them accessible to a wide audience, some images — such as the front panel of the TR-808 drum machine — will likely intimidate those who feel that the many knobs and

[*] To paraphrase D.H. Lawrence, who famously argued in his essay 'The Novel' that a novelist is a person who 'writes because he can't help it.'

buttons must be incredibly difficult to learn and operate. Other designs, such as the pianoforte, will appear more familiar, although in reality, a piano undoubtedly requires more practice than a drum machine to yield a satisfying result. But the most interesting tools are sometimes found lying on the kitchen table. Consider the pencil: centuries of design refinement and distilled knowledge — humanity's collective intelligence at its finest.

How we create largely determines *what* we create, which, when all put together, defines our culture. Hence, creative tools play a major role in shaping that culture. They don't appear by chance, or deterministically, or by strictly following scientific discoveries. Creative tools are *designed*: they are the fruit of decisions, both good and bad, made by humans. For now, at least.

In a society obsessed with consumerism and entertainment, designers have naturally developed a keener *interest* in products of *consumption* than in tools for *creation* — as attested by almost any popular design book. Logically, over time, it has become the norm to treat creation with sometimes ill-adapted ideas and models developed with the user-consumer in mind. If the negative effect on creation has so far been limited, it is partly due to the broad range of creative activities — each calling for distinct tools — and to the diverse approaches of those who design the tools that creators choose to use. However, this is changing.

While *creating* and *consuming* were clearly separate activities for most of human history, the 'softwarisation' of tools is blurring those boundaries. What used to be a wide range of tools is converging into one — whether a laptop or a smartphone — for which creative applications are designed with a single goal: attract an ever-growing 'community' of users and then drive them to push content to an even larger, ever-growing audience of consumers.

> *'How we create largely determines what we create, which . . . defines our culture.*
>
> *Creative tools play a major role in shaping that culture.'*

Modern creators are offered *automated* ways of doing things, designed to be *frictionless*: popular styles are easy to replicate, filters and effects are entertaining to apply, and results are guaranteed to ensure social validation.

And if it is tempting to dismiss this as an inconsequential area of human creation that can have only so much effect on culture at large, we should keep in mind that creation does not exist in a vacuum. The precarious life of the creator is an easy cliché *precisely* because it is a hard reality. Yet, if those who 'cannot not create' have so far found ways to make do, how are they going to fare when the only sustainable path forward is using tools inextricably bound with the monetisation of *content*? And how will those who don't comply fare when their art is put in front of an audience that simply has no *attention* left to spare?

Not that *content creators* have a guaranteed future either. The spectacular growth of the content economy, fueled by the rise of artificial intelligence, opens up a new paradigm where *creation* and *consumption* might merge entirely — the former absorbed by the latter until the only thing remaining of the complex nature of creativity is the vague momentary expression of desires. But, again, creative tools don't appear deterministically. Technology may enable more automation, but it does not *impose* it; humans who design *tools** that *prioritise* automation do — especially when they disregard creative agency to satisfy market growth imperatives. Hence, the challenge is not technological; it is ideological: the time and effort we spend on meaningless, dehumanised creation is the very mechanism by which we risk letting art, craft, and human creativity at large die, and the impact on our culture becomes all the worse.

Whatever the field, whatever the underlying technology, design always involves making choices: we can choose to design creative tools that embrace creative friction, seek diversity of expression, and leave ample opportunities for happy accidents along meaningful creative journeys. We can design tools that *truly* support human creativity — and in doing so, continue to enrich our culture.

'While technology expands the horizon of what is physically achievable, the design of tools proceeds from individual, granular, and often divergent decisions on how to apply that potential, if at all.'

---

[1] Sigmund Freud, *Leonardo da Vinci*, trans. A. A. Brill (1910; repr., n.p.: Acheron Press, 2018).

[2] D.H. Lawrence, 'The Novel,' in Phoenix: *The Posthumous Papers of D. H. Lawrence*, ed. Edward McDonald (London: Heinemann, 1936).

---

* The distinction between technology, as the practical application of scientific knowledge (as defined in the Introduction), and tools — specialised objects designed to facilitate a specific task — is crucial. While technology expands the horizon of what is physically achievable, the design of tools proceeds from individual, granular, and often divergent decisions on how to apply that potential, if at all.

As argued throughout Chapter 1, the risk of ignoring this 'granular power' is that we treat creators as mere consumers, prioritising an automated, frictionless experience that narrows the range of possible outcomes. Ultimately, it is through the tool that we reclaim intentionality, choosing whether to prioritise efficiency or to preserve the friction and limitations necessary for a genuine creative journey that allows for human agency and *happy accidents* (Chapter 4).

Image Labels & Credits

Page 7	HM King Charles III by Jonathan Yeo, Oil On Canvas (2024) — Courtesy of Jonathan Yeo
Page 33	Heydar Aliyev Center in Baku, Azerbaijan, by Zaha Hadid Architects (2012) — Photo by Evgeny Matveev
Page 54	LinnStrument MIDI Controller (2014) — Courtesy of Roger Linn Design
Page 57	Linn LM-1 Drum Computer Brochure (1980) — by Linn Electronics. Successor: Forat Electronics - Bruce Forat, president, Creative Commons Attribution-ShareAlike (CC BY SA 3.0 creativecommons.org) [Edited]
Page 60	reMarkable 1 Paper Tablet (2017) — Courtesy of reMarkable
Page 63	reMarkable Marker (2024) — Courtesy of reMarkable
Page 65	reMarkable Paper Pro Move (2025) — Courtesy of reMarkable
Page 67	Jonathan Yeo's Studio with 3D Painted Portrait Sculpture, Cast in Bronze (2017–2018) — Courtesy of Jonathan Yeo
Page 69	Bush by Jonathan Yeo, Collage on Board (2007) — Courtesy of Jonathan Yeo
Page 73	Writing for the Design Mind by Natalia Ilyin (2019) — Published by Bloomsbury Visual Arts
Page 74	Blog Post: I Am Not a Cruise Director by Natalia Ilyin (2015) — nataliailyin.net/blog/
Page 81	The Queen's Lyre, Partial Reconstruction (ca. 2500 BCE) — British Museum, London (commons.wikimedia.org/wiki/File:The_Queen%27s_Lyre,_side_view;_British_Museum,_London.jpg), Creative Commons Attribution (CC BY-SA 4.0 creativecommons.org) [Edited]
Page 81	Roman Pottery Workshop — Model by Andy Woodhead (sketchfab.com/Andywoodhead), Creative Commons Attribution (CC BY 4.0 creativecommons.org)
Pages 82-83	(Left) Assyrian Clay Cuneiform Tablet with a Small Second Tablet: Private Letter (ca. 20th–19th century BCE) — Metropolitan Museum of Art, Public Domain (CC0 1.0 Universal creativecommons.org) [Edited]; (Right) Wooden Block Movable Type — Wooden movable types with ligatures ft and fl by Raimond Spekking, Creative Commons Attribution (CC BY-SA 4.0 creativecommons.org) [Cropped]
Page 86	Gutenberg Press — Model by Daniel.Marquez2 (sketchfab.com/Daniel.Marquez2), Creative Commons Attribution (CC BY 4.0 creativecommons.org)
Page 87	Grand Piano by Bartolomeo Cristofori (1720) (Padua 1655-1731 Florence) — Metropolitan Museum of Art, Public Domain (CC0 1.0 Universal creativecommons.org) [Edited]
Pages 88-89	The History of the Artists' Colour Tube and Paint Tubes — Courtesy of Winsor & Newton and Colart
Pages 90-91	Singer Model 15 Sewing Machine — Photos by Olga Yastremska [Edited]
Pages 92-93	Underwood No. 5 Typewriter — Smithsonian National Museum of American History, Public Domain (CC0 1.0 Universal creativecommons.org) [Edited]
Pages 94-95	Leica 1 Camera — Courtesy of Leica Camera
Pages 98-99	RCA Theremin, Serial Number 100043 — Courtesy of Mike Buffington (RCATheremin.com)
Pages 100-101	Fender Stratocaster Guitar — Model by Ryan_Nein (sketchfab.com/Ryan_Nein), Creative Commons Attribution (CC BY 4.0 creativecommons.org) [Edited]
Page 104	Olivetti Lettera 22 Typewriter (1950) — Photo by Marcello Nizzoli (austincalhoon.com), Creative Commons Attribution (CC BY-SA 3.0, creativecommons.org) [Edited]
Page 104	Milling Machine — Model by LordSamueliSolo (sketchfab.com/LordSamueliSolo), Creative Commons Attribution CC BY 4.0 creativecommons.org)

Page 105	Minimoog Voyager Synthesiser used by J Dilla (2002–2005) — Smithsonian National Museum of African American History and Culture (CC0 1.0 Universal creativecommons.org)
Pages 106–107	(Left) Model House built of LEGO Bricks — Photo by Johan Nilsson Holmqvist; (Right) LEGO Bricks — Photo by Matt Hudson
Pages 108–109	(Left) Technics SL-1200 Record Player — Photo by pashapixel; (Right Top) Photo by ss_comm; (Right Bottom) Photo by hurricanehank
Pages 110–111	Korg MS-20 Synthesiser — Courtesy of Korg UK
Pages 112–113	Roland TR-808 Drum Machine
Pages 114–115	Yamaha DX7 Synthesiser — Courtesy of Yamaha Corporation
Pages 116–117	Wacom Intuos Pro Tablet and Pen
Page 120	Promotional Picture of Daft Punk after the release of Random Access Memories — by Sony Music Entertainment (commons.wikimedia.org/wiki/File:Daft_Punk_in_2013_2.jpg), Creative Commons Attribution (CC BY 4.0 creativecommons.org) [Cropped]
Pages 122–123	Akai MPC60 MIDI/Music Production Centre
Pages 124–125	Adobe Photoshop Screenshots
Pages 126–127	Blender Screenshots — Courtesy of Blender; Seiko Watch by carloshisserich (sketchfab.com/carloshisserich), Creative Commons Attribution (CC BY 4.0 creativecommons.org)
Pages 128–129	Unreal Engine Editor Screenshots — Courtesy of Epic Games
Pages 132–133	Pioneer CDJ-2000 Nexus — Photos by ARTYuSTUDIO [Edited]
Pages 134–135	Canon EOS 5D Mark II Camera — Courtesy of Canon
Page 136	DJ Playing Music for a Crowd in a Club — Photo by Laszlo Barta
Pages 138–139	Novation Launchpad Controller — Courtesy of Novation and Focusrite
Pages 140–141	Arduino UNO Board — Courtesy of Arduino, Arduino and UNO are trademarks or registered trademarks of Arduino S.r.l.
Page 143	Figma UI Component — Component by Joey Banks (figcomponents.com/kits/ios-15-ui-kit), Creative Commons Attribution (CC BY 4.0 creativecommons.org)
Pages 144–145	ARRI ALEXA Digital Cinema Camera — Courtesy of Arnold & Richter Cine Technik (ARRI)
Pages 146–147	Teenage Engineering OP-1 field Synthesiser — Courtesy of Teenage Engineering
Pages 148–149	MakerBot Replicator 2 3D Printer — Courtesy of MakerBot and UltiMaker
Page 150	Cinematic Shot of Mountains in Scotland — Photo by Suganth [Edited]
Pages 152–153	OpenAI DALL•E Generated Images
Pages 164–165	InSight Camera (2024) — by the authors/Product Design Studio Oh Hi Tomorrow
Page 166	Komanda Real-time Content Editing System (2017) — by the authors/Design Studio Wolf In Motion

Acknowledgements

We would like to extend our sincere thanks to Harm van Kessel and Peter Notebaart at BIS, whose support made this book possible.

We also warmly thank Roger Linn, Mats Herding Solberg, Jonathan Yeo, and Natalia Ilyin for generously taking the time to answer interview questions and share their experience, insights, and ideas on the design of creative tools.

We are grateful to the following individuals and organisations for generously responding to our requests with images that greatly enriched this book: Roger Linn and Roger Linn Design, Mats Herding Solberg and reMarkable, Jonathan Yeo, Ruben Cooke, and Elizabeth Duncan, Natalia Ilyin, Louise Baird-Smith, and Bloomsbury Visual Arts, Romain Guinier, Chris Wooff, Joe Davis, and Winsor & Newton/Colart, Analiese Oetting and the Smithsonian National Museum of American History, Mark Shipard, Stefan Daniel, Alexandra Althof, and Leica Camera, Mike Buffington, Luke Edwards and Korg UK, Kei Sato and Yamaha Corporation, Nicole Newman and inMusic/Akai Professional, Francesco Siddi and Blender, Jodie Anderson, Dana Cowley, and Epic Games, Eloise Pisano and Canon USA, Caitlyn G. Shaw and Gear Patrol/DPReview, Conor Boyd and Novation Music/Focusrite, Fabrizio Garda, Maria Tonini, Kateryna Lukina, and Arduino, Jonas Poetschke, Reegan Köster, Jan Heugel, and Arnold & Richter Cine Technik (ARRI), Emil Kullänger and Teenage Engineering, Andrea Zermeno, Felipe Castaneda, and MakerBot/UltiMaker.

Finally, this book would not have been the same without the thoughtful and incisive feedback of Esmeralda Atzamalidou, Govind Balakrishnan, Mike Buffington, Srikant Chakravarti, Gonzalo Herrero Delicado, Stephen Farrugia, Rodrigo García González, Markus Lahtinen, Juliette Shackleton, and Marie Tricaud.

Biographies

Guillaume Couche is the co-founder of London-based design studio Oh Hi Tomorrow, specialising in the design of creative tools. In 2013, he was part of the founding team of Gravity Sketch, a pioneering immersive 3D modelling application, and has since designed diverse devices for a wide range of creative fields, from music performance to sketching and filmmaking. In 2024, he was the co-recipient of a Red Dot: Best of the Best award for InSight, a futuristic photography camera. As a dilettante musician, film director and watercolourist, when he does not design creative tools, he uses them.

Sanya Rai is a Berlin-based Senior Interaction Designer at Google DeepMind who brings 15 years of experience in designing products across the physical-digital landscape, including Google Assistant, flagship Pixel phones and, most recently, the Gemini app. With a dual master's degree from the Royal College of Art and Imperial College London, she blends technical rigour with creative exploration. She previously held senior roles at Designit and has worked with organisations such as BMW, Huawei, Telekom and Nespresso. Approaching the design process as a vital creative journey, she applies UX, strategy, and design research to build beautiful user experiences that simply make sense.

Richard Shackleton is the co-founder of Oh Hi Tomorrow and has spent three decades shaping creative tools through product design, strategy, and innovation. His work has influenced technologies central to Hollywood film and animation, and in roles at Avid, Sony, Foundry, Allegorithmic, Epic Games, and others, he helped develop products used by studios such as ILM, Walt Disney Animation Studios, Wētā FX, Pixar, Marvel, and Netflix. He also supports emerging creative entrepreneurs as an investor and advisor. Outside the studio, Richard's interests include painting and scale model making, a reminder that the tools we create ultimately belong in the hands of makers.

Oh Hi Tomorrow is a product design studio in London: *Design isn't just about looks or functionality; it's about the story a product tells. We value experience and intuition over rigid process, asking critical questions to ensure each decision has a clear rationale and resonates with users. To bring product stories to life, we work hands-on to build physical models, code simulations, and apply our filmmaking skills—writing, directing, and editing—to convey the energy and emotion behind an idea. The studio holds many awards, including one for InSight*, a camera for spatial computing that won a Red Dot Design Concept: Best of the Best Award. Building on a foundation of theory and practice, we teach design at leading schools like the Royal College of Art, Imperial College London, and Lund University. They use our book, 'Interface Design: Creating Interactions That Drive Successful Product Adoption', which offers a unique conceptual approach to discussing and designing user interactions. We hope it, together with this book, challenges everyone to create innovative, engaging and successful future products.*

* See page 164.

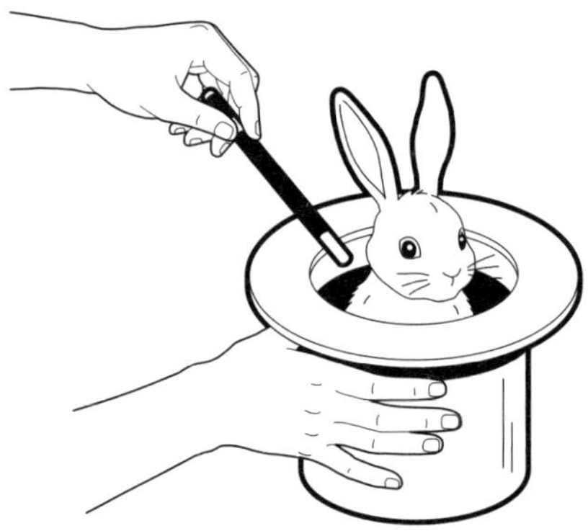